Shadow Artificial Intelligence

How Unsanctioned AI Became Your Biggest Blind Spot

Jazper Carter

Cybersoft Publishing LLC

Fort Washington, MD 20744

Drclaude.net

Second Edition March 2026

Contents

Introduction

This book is built around a paradox that every organization with more than a few hundred employees is facing. The workforce adopted artificial intelligence faster than any enterprise technology in history—not through a procurement process or a pilot program, but through individual decisions made thousands of times a day by people who found that a publicly available tool could do in seconds what used to take hours. The organization now has an AI footprint it did not plan for, cannot fully see, and has no governance structure to manage. The nine chapters ahead provide a systematic path from that starting condition to a functioning governance model.

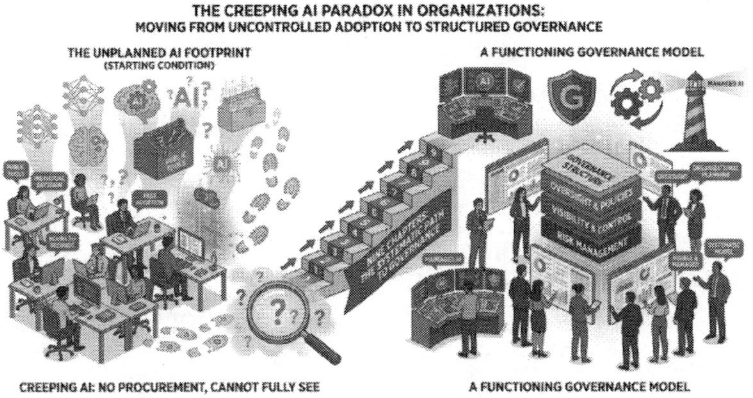

The first three chapters map the problem. You will begin with the mechanics of how shadow AI entered the enterprise, not as a single event, but as a convergence of accessible tools, remote-work norms, and procurement gaps that made unsanctioned adoption almost inevitable. The second chapter examines the productivity incentive from the employee perspective, because any governance model that ignores why people adopted these tools will fail to change the behavior. The third chapter catalogs the risk landscape data exposure, regulatory liability, decision integrity, and vendor lock-in, so that the governance response is proportional to the actual threat surface rather than driven by anxiety.

Chapters four through six shift from diagnosis to architecture. The discovery chapter addresses the hardest operational question in the entire book: how do you inventory tools that were adopted specifically to bypass enterprise procurement and leave minimal footprints? From there, you will design the governance framework itself, policy structures that people will follow rather than circumvent, and establish the shared accountability model that distributes ownership across IT, legal, compliance, human resources, and executive leadership.

The final three chapters are about conversion and sustainability. You will find a structured methodology for bringing unsanctioned tools under governance without shutting down the productivity gains they enabled, followed by the organizational pivot from a prohibition posture to a responsible enablement model. The closing chapter addresses the hardest long-term challenge: maintaining governance visibility as the AI tool landscape continues to evolve faster than any policy review cycle can match.

This is not a book that requires technical depth to use. It is written for the leader whose responsibility spans the organizational response. This executive needs to coordinate across functions, allocate resources, set policy direction, and measure whether the governance model is working. Read it sequentially for the first time to build the full decision framework. After that, return to individual chapters as operational needs arise. Each one is designed to function as a standalone reference for the specific governance challenge it addresses.

1 The Invisible Deployment

1.1 A Discovery No One Planned

The security operations team at a mid-sized financial services firm received an alert on a Tuesday morning that stopped several conversations mid-sentence. An automated data loss prevention scan had flagged outbound traffic to three AI model endpoints that appeared on no approved vendor list, had no firewall policy exception, and had no procurement record. When investigators traced the traffic back to its source, they found that dozens of employees across four departments, loan origination, compliance, marketing, and legal, regularly sent structured data to consumer large language model services. None of those employees had asked for permission. Most assumed someone else had cleared it. A few assumed the absence of a prohibition meant implicit approval. None of them was acting in bad faith. Everyone was trying to do their job faster.

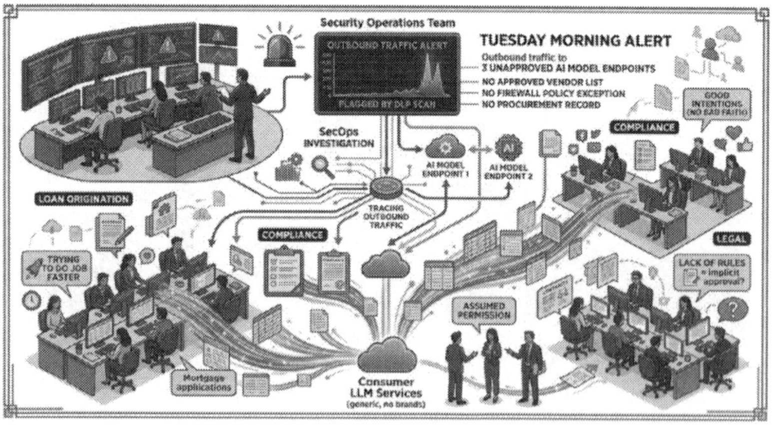

That incident, reconstructed from accounts of similar events across multiple industries, captures the defining feature of shadow AI in the modern enterprise: it is not a technology problem that arrived from the outside. It is a behavior pattern that grew from the inside, quietly, in the space between what employees needed to accomplish and what the organization had provided to help them. By the time a security team notices it, the practice is already embedded in workflows, relied upon, and, in some cases, entangled with consequential decisions.

The first step toward governing shadow AI is to understand how it entered your organization without a signed agreement, a security review, or an approved budget line. That entry was not an accident. It was the predictable consequence of structural conditions that most enterprises created or failed to prevent well before the current generation of AI tools became accessible to every employee with a browser.

1.2 Why This Chapter Matters

Most governance conversations begin at the wrong point. They open with a policy on what should be allowed, who should approve it, and what enforcement mechanisms should be in place. Those are important questions, but they come too late in the analysis if you have not first understood how AI entered the organization and why the entry conditions remain in place. If you design governance for a problem you have misdiagnosed, your policies will address the surface behavior while leaving the structural conditions intact. Employees will find new entry points, and the cycle will repeat.

This chapter traces the mechanics of invisible deployment across three interrelated dimensions: the procurement and detection gaps that allowed AI tools to reach enterprise workflows without evaluation; the transformation of individual productivity habits into organizational data flows; and the reasons traditional IT governance arrived too late to intercept shadow AI before it became operationally embedded. The goal is operational clarity, a precise map of how the problem formed so that your governance response addresses the actual terrain, not a simplified version of it.

Shadow AI carries workflow-level impact in ways that earlier shadow IT did not. When an employee used an unsanctioned cloud storage tool a decade ago, the primary risks were data residency and access control. Those risks were real, but they were bound. When an employee uses an unsanctioned large language model today, they may be transferring proprietary data into a training pipeline, generating outputs that inform hiring decisions, producing text that enters customer communications, or creating outputs that constitute legal advice, none of which is reviewed, attributed, or reversible. The exposure surface is categorically larger, and the governance response must match that scale.

Diagram 1.1: The Shadow AI Entry Map

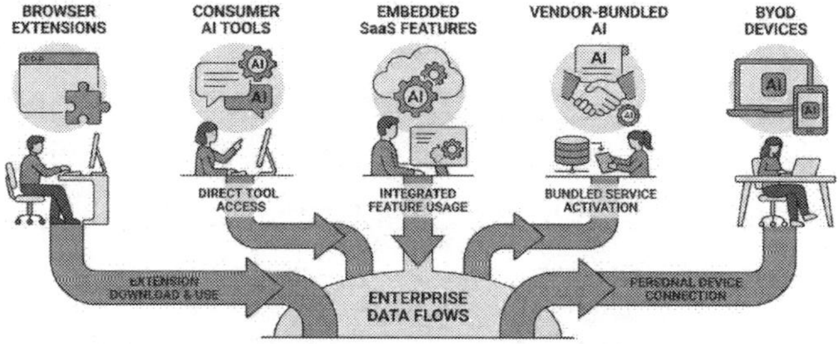

Unmonitored and unauthorized AI adoption introduces data leakage risks through five key enterprise entry pathways.

1.3 From Shadow IT to Shadow Intelligence

The category of "shadow IT" technology employees use without formal approval has existed as long as enterprise IT governance has. The term gained traction in the early 2000s as consumer software became sophisticated enough to compete with enterprise applications, and employees began routing work through personal Gmail accounts, Dropbox folders, and consumer productivity tools rather than waiting for IT to provision approved equivalents. Shadow IT was a management headache, a compliance risk, and an indicator of unmet employee need. But it was, in most cases, manageable.

Shadow AI is different in kind, not just in degree. The distinction matters because it determines the appropriate response to governance. Earlier shadow IT primarily created access-control problems: files in unsanctioned locations, communication in unmonitored channels, and software with unreviewed security postures. The remedy was largely structural: better procurement processes, faster tool provisioning, clearer acceptable-use standards, and technical controls that enforced those standards at the network layer. Shadow AI creates all of those problems and adds several that have no direct precedent in shadow IT history.

The first new problem is data transfer at the point of use. When an employee pastes a customer contract, a patient record summary, a draft earnings report, or a personnel performance review into a large language model prompt, they are not accessing an unsanctioned file-sharing service. They are transmitting organizational data to a third-party system whose data retention and training data policies may not be consistent with the organization's regulatory obligations, contractual commitments, or competitive security requirements. The transfer occurs at the moment of use, leaves no artifacts in enterprise systems, and, in many implementations, cannot be recalled.

The second new problem is decision implication. Large language models do not passively store files; they generate outputs that employees use to make decisions, draft communications, evaluate options, and summarize complex material. When those outputs are produced by unsanctioned tools operating without organizational oversight, the decisions they inform carry organizational weight without organizational attribution. The manager who used an AI tool to rank job applicants, the analyst who used one to summarize a regulatory filing, and the customer service representative who used one to draft a response to a client complaint are all making consequential decisions through a system the organization has not reviewed, approved, or documented.

The Procurement Blind Spot

Enterprise procurement processes were designed for a different technological era. The implicit model was that significant software capabilities required significant procurement: a contract, a legal review, a security assessment, a license agreement, and a deployment project. That model worked reasonably well when enterprise software was expensive, complex to deploy, and required organizational infrastructure to operate. It performs poorly when capable AI tools are available in a browser at no cost or bundled into subscription services that employees already use.

The procurement blind spot has three components. The first is the free-tier problem: many of the most capable consumer AI tools are available at no cost, so they generate no expense reports, purchase orders, or budget line items. Procurement processes that rely on financial signals to trigger reviews will never use these tools. The second component is the bundling problem: AI capabilities are increasingly embedded in software products that organizations have already purchased and approved. When a project management tool adds an AI-assisted writing feature, or when a customer relationship management platform adds AI-generated call summaries, the AI capability enters the enterprise under the cover of the original procurement decision. The third component is the velocity problem: AI tool capabilities are evolving faster than enterprise procurement cycles can accommodate. A tool that was reviewed and approved twelve months ago may now include capabilities that were not present at the time of review, and that would have triggered additional scrutiny.

THE PROCUREMENT BLIND SPOT: THREE COMPONENTS

Closing the procurement blind spot requires governance mechanisms that do not rely on financial signals, include AI capability reviews in SaaS renewal processes, and treat AI feature updates from approved vendors as potential governance triggers. The mechanics of each of these responses are addressed in later chapters; the point here is that the blind spot is structural, not incidental, and will not be addressed by awareness campaigns or policy statements alone.

Consumer-Grade Models at Enterprise Scale

The consumerization of AI has followed a trajectory that enterprise governance was poorly positioned to anticipate. In previous technology transitions, such as personal computers, mobile devices, and cloud storage, there was a recognizable period during which consumer tools were clearly less capable than enterprise alternatives, giving organizations time to evaluate and provision sanctioned options before employee adoption reached scale. That transition period is largely absent from the current AI wave.

The rate at which AI model capabilities have improved has also outpaced the typical enterprise vendor evaluation cycle. Security assessments and procurement reviews evaluate a product at a point in time. Still, AI model capabilities evolve continuously through retraining, fine-tuning, and architecture updates that may not trigger a formal product release. An organization that approved an AI writing assistant based on a review twelve months ago may now have deployed a capability substantially different from what was evaluated, and neither the organization nor the vendor has necessarily flagged the change as requiring supplemental review. Building governance that accounts for capability drift, not just the initial product, but its ongoing evolution, is one of the more operationally difficult challenges in enterprise AI management.

Consumer large language models available today routinely outperform enterprise-specific tools in general reasoning, language generation, and analytical capability. An employee comparing an enterprise-approved document summarization tool with a consumer AI service is, in many cases, making an accurate assessment when they choose the consumer option. The governance challenge is not correcting a misperception about tool quality; it is building compliance frameworks that acknowledge employee productivity needs while addressing the risk dimensions that individual users are not equipped to evaluate.

Consumer-grade models operating at enterprise scale create a specific risk profile. These tools were designed for individual users, not for enterprise environments. Their default data retention policies, training data practices, logging configurations, and access controls are calibrated for consumer use cases, not for environments governed by regulatory frameworks such as HIPAA, FedRAMP, SOX, or GDPR. When employees use these tools for enterprise work at scale, they are importing consumer-grade data governance into an enterprise-grade risk environment. That mismatch is the core operational problem that this book addresses.

Diagram 1.2: Consumer AI vs. Enterprise Risk Environment

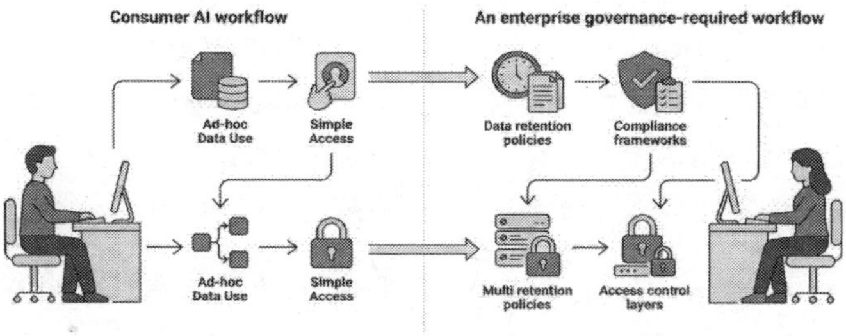

Consumer AI defaults create substantial gaps that robust enterprise governance frameworks must address for safe and compliant adoption.

1.4 The Browser Extension Problem

If the procurement blind spot represents AI entering through organizational gaps, the browser extension problem represents AI entering through the front door embedded in tools that employees use every day, enabled by default or with a single click, and operating invisibly within approved workflows. Browser extensions have become one of the primary vectors for the adoption of shadow AI in enterprise environments, precisely because their deployment model bypasses every governance checkpoint that IT departments have traditionally relied on.

The modern browser is a full enterprise application platform. Employees conduct procurement workflows, access financial systems, communicate with clients, and manage human resources processes through browser-based interfaces. In many implementations, browser extensions installed on those same devices can read page content, access form inputs, intercept network traffic, and inject content into pages. An AI-enabled browser extension installed by an employee on an enterprise device can, depending on its permissions, observe every web-based workflow that the employee conducts. The governance implications are significant, whether the extension is malicious or benign.

Embedded and Invisible: AI in Approved Software

The more consequential category of embedded AI is not malicious extensions, but legitimate AI features built into products that the organization has already approved. Enterprise software vendors across every category, productivity suites, customer relationship management platforms, project management tools, communication applications, and development environments, have added AI capabilities to their products in ways that organizations did not evaluate when they made the original procurement decision.

These embedded features pose a specific governance challenge: they are technically within the scope of the approved product, but they introduce data-handling behaviors that were not present when the product was evaluated. When a project management platform adds an AI feature that summarizes task comments and recommends resource allocations, the data flowing through that feature may be subject to different retention policies, different third-party sharing arrangements, and different model training practices than the rest of the platform. The original procurement review did not address those behaviors because they did not exist at the time of review.

The responsible-by-design response to embedded AI features is to treat significant additions to AI capabilities as triggering events for supplemental vendor review. This requires procurement and vendor management processes that monitor product release notes for AI capability announcements and that have a defined pathway for evaluating new features in already-approved products. In practice, most organizations do not have this mechanism in place, which means embedded AI features accumulate in the enterprise tool stack without review until a security or compliance event forces the issue.

Copilot Toggles and Default-On Features

Microsoft 365 Copilot, GitHub Copilot, Salesforce Einstein, Google Workspace AI features, and dozens of similar enterprise AI integrations share a common deployment pattern: they are enabled at the platform level, often with organizational administrators able to turn them on or off, but individual users can frequently enable or configure them at the account level within whatever permissions the platform administrator has set. The result is a governance environment where the AI capability is technically within the approved product boundary. Still, its activation and configuration decisions are distributed across thousands of individual users rather than centrally managed.

The default-on pattern compounds the toggle problem. Many vendors have moved toward enabling AI features by default in new releases, requiring administrators to opt out rather than opt in explicitly. Organizations that do not actively monitor software updates and evaluate default configurations will find AI features enabled across their environment without a conscious governance decision. The absence of an opt-out is not the same as an informed opt-in, but in many enterprise environments, it functions that way.

For managers, the operational implication is that the question "what AI tools are my team using?" cannot be answered by looking at purchase orders or approved tool lists. It requires an inventory process that examines what features are enabled within approved platforms, what default configurations are in place, and what user-level configuration options are available. That inventory process is addressed in Chapter 4; the point here is that the boundary between sanctioned and unsanctioned AI is blurrier than most managers assume.

SaaS Platform Updates as Stealth AI Rollouts

Software-as-a-service platforms update continuously and without the formal change management cycles that on-premise software required. In an on-premise deployment, a software update required organizational testing, a deployment window, and administrator action. In a SaaS environment, features can appear in production overnight, visible to users before IT or management is aware of their existence. Vendors may announce AI features in release notes, blog posts, or developer documentation, which enterprise governance teams rarely monitor at the frequency required to intercept capabilities before employee adoption begins.

The practical effect of continuous SaaS updates is that the AI capability landscape within an enterprise changes constantly and often without explicit organizational awareness. A contract management platform that did not have AI features in January may have AI-assisted clause analysis, AI-generated contract summaries, and AI-recommended redlines by March. Employees who discover these features in the natural course of their work will adopt them if they find them useful, which they typically do. By the time governance processes recognize the new capability, it may already be embedded in standard workflows.

Diagram 1.3: SaaS AI Feature Accumulation Over Time

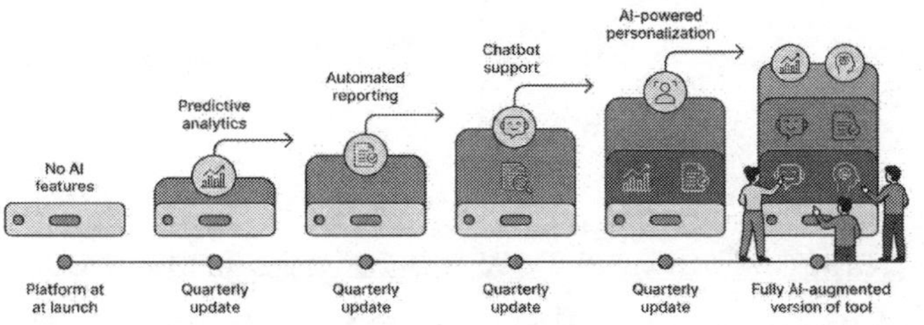

Strategic quarterly releases progressively integrate specialized AI features, building a more powerful and efficient SaaS tool over time.

1.5 Anatomy of an Unsanctioned AI Workflow

Understanding shadow AI at the individual tool level, which products employees are using, and which features are enabled, provides an incomplete picture. The more important unit of analysis is the workflow: the sequence of steps through which work gets done, the data inputs those steps require, and the outputs they produce. Shadow AI becomes an enterprise governance problem not when an individual employee experiments with a tool once, but when that individual uses it, it hardens into a repeatable workflow, and that workflow diffuses across a team, a department, or an organization.

Tracing the anatomy of an unsanctioned AI workflow reveals how quickly individual behavior can create structural organizational risk. The progression from individual workaround to embedded organizational practice typically follows a recognizable pattern, and understanding that pattern gives managers the analytical vocabulary to intervene at the right point in the sequence.

How a Single Productivity Habit Becomes an Enterprise Data Flow

The pathway from individual experimentation to organizational data flow typically begins with a moment of friction, a task that takes longer than it should, a capability gap in an approved tool, or a workflow that feels manual in a way that a consumer AI tool can obviously address. The employee who discovers that a large language model can condense a thirty-page vendor proposal into a three-paragraph executive summary in under a minute is making a rational productivity decision when they begin to do so regularly.

The next stage is informal endorsement. The employee mentions the approach to a colleague. The colleague tries it. The team lead, who is measured on throughput, observes faster output and asks no questions. Within a few weeks, what began as individual experimentation is team practice. The data flowing through the tool has expanded from one employee's interactions to a team's worth of vendor proposals, client documents, internal reports, and strategic memos.

The final stage is normative embedding. The tool becomes part of how work is described: "run it through the AI first, then we'll review." New team members are onboarded into the practice as part of informal knowledge transfer. The workflow is now organizational, but it exists entirely outside any governance structure. The procurement process never evaluated it. IT has no visibility into it. Legal has not reviewed the vendor's data handling terms. Compliance has not assessed whether the data flowing through it meets regulatory handling requirements. The individual productivity decision has become an organizational data governance problem.

From Individual Workaround to Team Norm

The diffusion of shadow AI from individual to team to organizational norm follows social dynamics that are well understood in organizational behavior research, even if they are underappreciated in technology governance contexts. Employees do not treat AI tool recommendations the way they treat technical recommendations from IT: they treat them as productivity peer advice, which travels faster, carries more credibility with immediate peers, and encounters less friction during adoption than formal tool provisioning.

The team norm problem is particularly acute because it creates a social pressure dimension that pure technical controls cannot address. When a tool is embedded in team norms, turning it off does not simply remove a capability; it disrupts established work patterns, creates perceived productivity regression, and generates friction in employee relations. Managers who discover that their team has been using unsanctioned AI and must now remove it face not just a technical removal task but a change management challenge. The governance implication is that early-stage intervention when individual use is still just individual use is dramatically less costly than late-stage intervention after team norms have formed.

Diagram 1.4: Workaround to Organizational Norm Progression

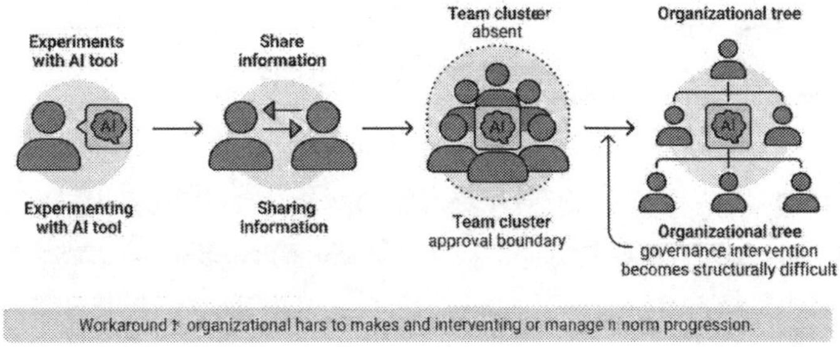

Identifying the AI Surface Area

The "AI surface area" of an enterprise is the aggregate of AI touchpoints through which organizational data flows or AI-generated outputs influence organizational decisions. It includes sanctioned tools, unsanctioned tools, embedded features in approved platforms, and AI capabilities in vendor-operated systems that process organizational data on the enterprise's behalf. Most organizations have not mapped this surface area, so their governance posture is defined by the portion they know about rather than the full extent of their exposure.

Mapping the AI surface area is the starting point for any credible governance program. The inventory process required to do this is detailed in Chapter 4. Still, the conceptual frame is important here: the surface area is almost certainly larger than your current tool list suggests; it is growing continuously as vendors add AI features to approved products; and it includes both intentional AI adoption and AI capabilities that are activated through vendor product decisions rather than organizational choices. A governance program that addresses only intentional adoption is addressing a fraction of the actual surface area.

For managers, operational clarity about the AI surface area means being able to answer three questions at any point: what AI systems does my team interact with, what data flows through those systems, and what decisions are informed by those systems' outputs? The answer to none of these questions should be "I don't know," because not knowing is not a neutral governance posture; it is an exposure posture. The chapters that follow provide the tools to move from exposure to oversight.

1.6 Why Traditional Governance Arrives Too Late

Traditional IT governance was designed for a technology adoption cycle that no longer describes how enterprise AI enters organizations. The classic governance model assumed that significant technology capabilities required significant organizational commitment to deploy infrastructure, budget, training, and vendor negotiation, which created natural checkpoints at which governance could be applied. The employee experience of enterprise AI today involves none of those checkpoints. A capable AI tool is available in a browser tab, functional within seconds of account creation, and requires no organizational infrastructure to operate at the individual level. By the time traditional governance mechanisms engage, adoption may be widespread and embedded.

The mismatch between governance design and adoption mechanics creates what might be called the policy-deployment gap: the interval between when a technology becomes widely available to employees and when organizational governance formally addresses it. In previous technology generations, this gap was measured in months and could be closed through reasonably responsive procurement and policy processes. In the current AI generation, capable tools can reach millions of individual users within days of release, and the gap can be measured in weeks, with organizational adoption preceding governance by a significant margin.

The Policy-Deployment Gap

The policy-deployment gap has a specific operational signature. When governance teams discover a shadow AI practice that has been running for months, they typically find not just a tool in use, but a set of workflow dependencies, informal documentation practices, and team norms built around that tool. Removing or restricting the tool at that point incurs real operational costs: work that was being completed faster must now be completed more slowly or with different tools, and the productivity regression is felt immediately, while the risk reduction from the restriction is abstract and distant.

This asymmetry, the concrete near-term cost of restriction versus abstract long-term risk reduction, shapes how operational managers respond to governance interventions, even when those managers understand the rationale. It is one of the core reasons that shadow AI governance requires more than policy: it requires a pathway for addressing legitimate employee productivity needs through sanctioned channels at a speed that can compete with uncontrolled adoption. The governance posture described in this book is not primarily prohibitive; it is a response design that gives employees faster access to evaluated tools, thereby reducing the incentive to route around governance.

Risk Tolerance by Seniority

Shadow AI adoption is not evenly distributed by seniority level, and the distribution pattern has significant governance implications. Individual contributors and mid-level professionals tend to adopt AI tools to improve task-level efficiency, writing assistance, summarization, data formatting, and research acceleration. Their risk tolerance is calibrated to the immediate task: if the tool helps and no adverse consequences materialize, the implicit risk assessment is favorable. Senior leaders, paradoxically, often exhibit higher-impact shadow AI behavior: using AI tools to prepare for board presentations, synthesize competitive intelligence, draft communications to investors or regulators, and inform strategic decisions. The organizational weight of these decisions is substantially higher than individual contributor use cases, and the governance gap is more consequential.

The risk-tolerance dynamic poses an awkward governance challenge. Governance programs often focus enforcement attention on rank-and-file employees because they are more numerous and their behavior is more visible in network logs. Senior leader AI use is less visible, less frequently flagged by technical controls, and often subject to informal norms that treat executive productivity tools as matters of personal discretion rather than organizational policy. A mission-aligned governance program must address the full seniority spectrum with appropriate controls and with leadership behavior that models the compliance standard it expects from others.

Speed as the Enemy of Oversight

The AI capability release cycle operates on a timeline that is structurally incompatible with traditional governance throughput. Enterprise security reviews, legal assessments, and procurement evaluations are calibrated to thoroughly evaluate capabilities that will be in place for months or years. They are not calibrated for a technology landscape in which significant new capabilities are released weekly, in which capability characteristics change through model updates without formal product release cycles, and in which competitive pressure on vendors creates incentives to deploy capabilities before enterprise governance frameworks have addressed them.

Speed is not merely an inconvenience for governance; it is a structural adversary. The governance response is not to attempt to match the velocity of AI capability releases, as race cannot be won with traditional governance mechanisms. The response is to build governance frameworks that operate under conditions of incomplete information, use risk-tiered evaluation to reserve deep review for high-consequence use cases, and provide fast-track pathways for lower-risk tools. The tiered governance approach is detailed in Chapter 5; the point here is that governance designed for a different technology pace will fail not because its principles are wrong but because its throughput is insufficient.

Diagram 1.5: The Policy-Deployment Gap

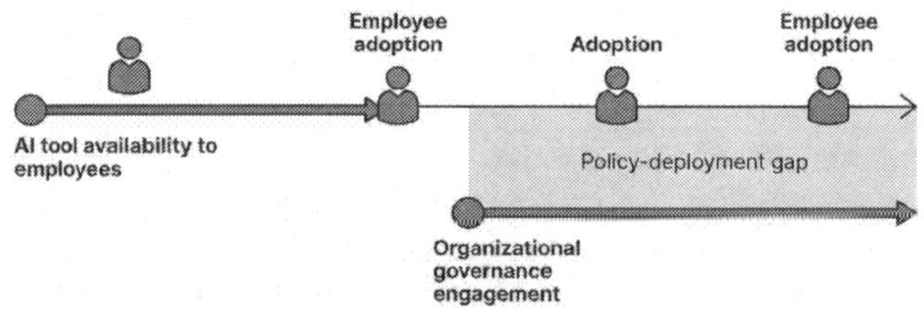

The policy-deployment gap visualizes the dangerous period between rapid AI tool availability and delayed organizational governance implementation.

1.7 Manager's Checklist: Chapter 1

The following behaviors and artifacts represent the minimum level of governance awareness a manager should establish before moving on to subsequent chapters. This checklist is diagnostic, not exhaustive. It identifies the gaps most commonly found in organizations that have not yet systematically addressed shadow AI.

1. Conduct an informal team conversation about what AI tools individuals are using, including consumer tools, browser extensions, and AI features in approved platforms.
2. Review the AI-related terms in the data handling and acceptable use agreements for the three SaaS platforms your team uses most frequently.
3. Identify whether your organization has a formal AI acceptable use policy and whether your team has been trained on it within the past twelve months.
4. Ask your IT or procurement contact whether the organization has a mechanism for reviewing AI feature additions to approved SaaS products.
5. Assess whether your team's performance metrics create implicit incentives for speed that might motivate shadow AI adoption.
6. Determine whether any AI-generated output currently influences decisions in your team's workflows and whether those outputs are documented or attributed.
7. Review whether your team has received guidance on what categories of organizational data should not be entered into external AI tools under any circumstances.
8. Identify one workflow in your team's current practice where unsanctioned AI use is most likely to have emerged and investigate whether that assessment is accurate.

1.8 Key Takeaways

Shadow AI is not a technology anomaly; it is the predictable outcome of a structural mismatch between employee productivity needs, the availability of capable consumer tools, and governance frameworks designed for a slower technology adoption cycle. By the time most organizations recognize they have a shadow AI problem, the problem is already embedded in workflows, normalized in team culture, and generating ongoing exposure that accrues daily.

The invisible deployment of AI into enterprise environments happened through multiple simultaneous channels: direct consumer tool adoption, browser extensions, embedded features in approved SaaS platforms, and AI capabilities bundled into existing vendor relationships. No single detection or governance mechanism addresses all of these channels, which is why effective shadow AI governance requires a system rather than a policy.

The foundational insight of this chapter is that shadow AI is a governance design problem, not a compliance enforcement problem. Organizations that approach it as an enforcement challenge will be continually outpaced by adoption. Organizations that approach it as a design challenge, asking how governance can be structured to address legitimate employee needs while managing actual risk, will build programs that are sustainable, adopted, and mission-aligned. The chapters that follow build that design, starting with a deeper examination of why employees choose unsanctioned tools in the first place.

Diagram 1.6: Shadow AI Governance System Overview

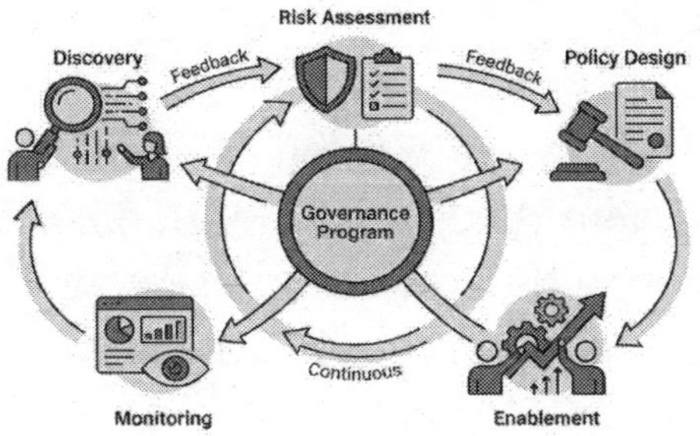

A continuous feedback loop to identify and manage shadow AI across the organization.

2 The Productivity Imperative

2.1 A Decision Made Before Lunch

A policy analyst at a federal agency sat in front of a deadline on a Wednesday morning: a summary brief on three competing regulatory frameworks was due by the end of business, and the approved research tools she had access to through the agency portal would have taken her the better part of the day. Instead, she opened a browser tab, navigated to a consumer large language model she had been using for three months, and completed a working draft within ninety minutes. She added her analysis, verified the framework descriptions against primary sources, and submitted the brief on time. No information classified above the basic threshold entered the prompt. No personal data was transmitted. The output was reviewed and edited by a subject matter expert before it left her desk.

By every operational measure, the interaction was productive, and the output was sound. By every measure of governance, the interaction was a policy violation. She had routed government work through an unapproved, unassessed, consumer AI system operating under data retention terms that the agency had never reviewed. The gap between those two assessments, operationally defensible, governance non-compliant, is where shadow AI lives. It is not the gap between good intentions and bad behavior. It is the gap between approved capabilities and actual workflow needs, and it will not close until organizations address both sides of the equation.

Understanding why shadow AI proliferates requires looking past the policy violation and into the experience of someone trying to do their job well in an environment where the approved toolkit is consistently behind the available technology. That experience, repeated across millions of workers in enterprises of every size and sector, is the structural driver of shadow AI adoption. It cannot be managed out of existence by enforcement alone. It must be understood well enough to design governance responses that address the underlying conditions, not just the surface behavior.

2.2 Why This Chapter Matters

Governance programs that treat shadow AI adoption as primarily a compliance failure will invest their energy in detection and deterrence. Those are legitimate governance functions, but they address the symptom rather than the cause. The cause is a structural misalignment between what employees need to do their jobs effectively and what the organization provides to support that. Until that misalignment is addressed, deterrence will be a rearguard action continuously fighting adoption pressure with enforcement tools while the underlying conditions that generate adoption remain in place.

This chapter examines the structural and psychological drivers of the adoption of shadow AI in enterprise environments. It maps the enablement gap, the interval between what approved tools offer and what available tools offer. It analyzes how productivity pressure, manager behavior, and performance measurement systems shape individual decisions to route work through unsanctioned AI. It then examines constraint environments, particularly the public sector and healthcare, where formal compliance requirements are highest, and workaround behavior is most sophisticated. The chapter closes with a framework for translating observed unauthorized behavior into organizational requirements, converting a governance liability into a product design input.

The mission-aligned governance response to productivity-driven shadow AI is not to accept compliance violations for productivity reasons. It is to build governance systems that can move fast enough to address legitimate productivity needs through sanctioned channels, so the incentive to route around governance is minimized at the source. That requires understanding the actual need, which this chapter's analytical work enables.

Diagram 2.1: The Enablement Gap Model

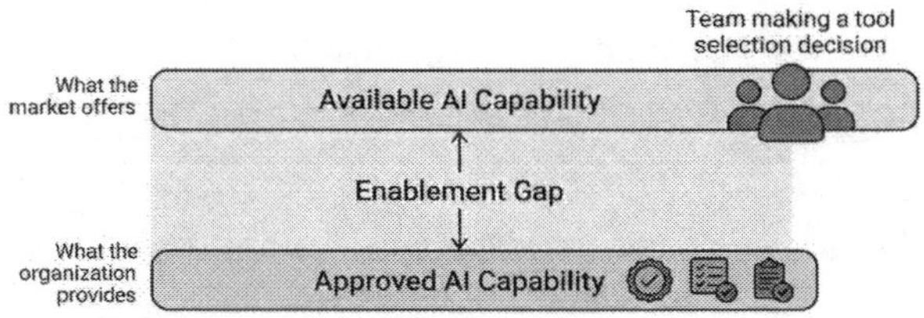

Visualizes the gap between market AI capabilities and internally approved tools.

2.3 The Enablement Gap

The enablement gap is the distance, at any given moment, between what technology the market makes available and what technology the organization has evaluated, approved, and provisioned for employee use. This gap has always existed to some degree; enterprise procurement has never kept pace with technology development. But the gap in the current AI period is qualitatively different from historical technology gaps, because the capability difference between what is available and what is approved is large, the tools on the available side require no organizational infrastructure to operate, and the productivity advantage they offer is immediately visible to the individual employee.

In previous technology transitions, bridging the enablement gap required organizational action: procuring licenses, deploying infrastructure, and training users. The friction of individual tool adoption was high enough that most employees waited for organizational provisioning. AI tools available today require none of that. An employee can create a free account, begin a productive interaction, and observe a clear capability advantage within minutes. The friction of individual adoption is near zero, which means the enablement gap translates directly into adoption pressure without the natural friction that historically gave governance time to respond.

Closing the enablement gap requires governance mechanisms that can evaluate and provision AI tools at a pace that competes with individual adoption rates. That does not mean evaluating every tool with the same depth; it means building a tiered evaluation framework that matches evaluation rigor to risk level, so that lower-risk tools can be provisioned quickly enough to meet productivity demands before employees turn to unsanctioned alternatives. The tiered framework is detailed in Chapter 7; the enablement gap concept here establishes why speed of governance response is itself a governance design variable.

Tool Request Backlogs and the Workaround Decision Tree

In most enterprise environments, the formal pathway for requesting a new software tool involves a request form, a review queue, a security assessment, a legal review of terms of service, a budget approval, and a deployment process. The elapsed time from initial request to available tool varies widely by organization, but industry surveys consistently show median approval timelines measured in weeks rather than days. An employee facing an immediate productivity need will not wait weeks for a formal approval process. They will make a workaround decision.

The workaround decision tree is not random. Employees evaluate available options using an informal risk calculus: Is this tool free? Does it require me to enter any sensitive information? Will anyone know I am using it? Has anyone else on the team used it? The consumer AI tools that win this informal evaluation are those with high perceived utility, low visible risk, and social proof from peers. Governance programs that want to intercept the workaround decision must compete at this level, providing faster formal pathways so that the approved option appears before the workaround decision is made.

Some organizations have introduced an AI tool that offers fast-track processes, committing to a preliminary assessment within five business days and a full evaluation within thirty business days. Those timelines are still long enough to miss the moment of immediate need, but they dramatically narrow the window in which employees resort to workarounds without a viable formal alternative. The operational goal is not to eliminate workaround decisions through prohibition; it is to make the formal pathway faster and more visible than the informal one.

Benchmarking the Approval Lag

Organizations that have not measured their tool approval lag typically underestimate it significantly. The relevant metric is not the stated policy timeline, the number of days the process is supposed to take, but the actual elapsed time from employee request to available tool, measured across a representative sample of recent requests. Most organizations that have conducted this measurement discover that their actual lag substantially exceeds their stated policy, because review queues, organizational dependencies, and handoff delays are not visible in the policy document but materialize in practice.

Benchmarking the approval lag against the AI tool adoption velocity in the market produces a concrete gap estimate. If the organization's approval process takes six weeks on average and major AI tools reach widespread adoption within days of release, the gap is approximately 40 days during which employees make tool decisions without a viable formal pathway. That gap is not a compliance failure; it is a governance design failure, and it can be addressed through process redesign rather than enforcement escalation.

The benchmark conversation is also useful for surfacing where the bottlenecks actually are. Security review, legal terms assessment, and budget approval are common chokepoints, and each has different remediation options. Pre-approved vendor security frameworks, standard AI tool terms assessment checklists, and delegated budget authority for low-cost tools are examples of structural changes that reduce approval lag without reducing review quality. Measuring the current state is the prerequisite to designing the improved state.

The Approved-Tool Graveyard

Every enterprise has a version of the approved-tool graveyard: a catalog of software that has been formally evaluated, approved, licensed, and provisioned, but that employees do not use because the tools are slower, less capable, or more cumbersome than available alternatives. The approved-tool graveyard is significant for shadow AI governance because it represents a credibility problem: employees who have experienced the graveyard are rational skeptics of governance claims that approved tools meet their needs. When governance tells employees that a sanctioned AI tool is available, employees who remember previous approved tool disappointments apply that experience as a prior.

Addressing the approved-tool graveyard is partly a procurement problem, selecting better tools, and partly a communication problem, surfacing the governance investments that capability investments have made. But it is primarily a responsiveness problem: governance programs that consistently provision tools that fall short of employee expectations lose the trust necessary to make governance work. A governance program that acknowledges capability gaps, explains why certain tools have limitations, and demonstrates a track record of improving provisioned capabilities over time is more likely to earn employee cooperation than one that insists on tool adequacy that the workforce has already evaluated and rejected.

Diagram 2.2: Tool Request to Deployment Timeline

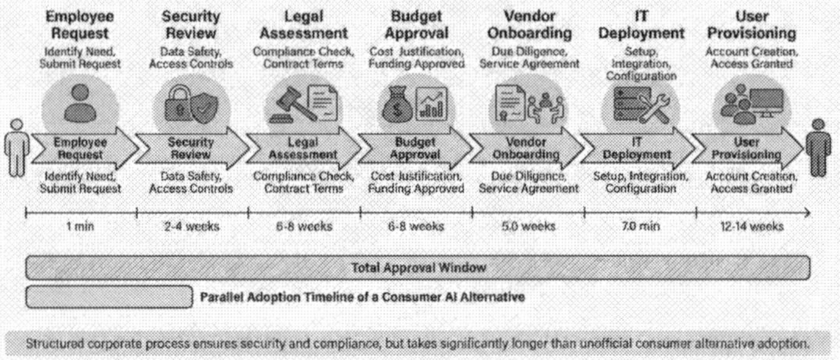

2.4 Productivity Pressure as Governance Risk

Productivity pressure is not simply a motivation for individual employees; it is an organizational structure that shapes governance-relevant behavior across entire teams and departments. When organizations set output metrics, reward speed, and evaluate managers on throughput without explicitly accounting for governance compliance in those evaluations, they create a structural incentive for workaround behaviors that produce shadow AI. Understanding productivity pressure as a governance risk factor, rather than as an individual motivation, is essential for designing governance responses that address its systemic dimensions.

The relationship between productivity pressure and shadow AI adoption is not merely correlational. In environments where output velocity is the primary visible performance metric, employees who use faster tools, including unsanctioned AI, will produce more visible output than employees who comply with governance constraints that slow their work. Over time, this creates a compliance penalty: compliant employees appear less productive than non-compliant ones, which creates a selection pressure against compliance. Governance programs that ignore this dynamic will be fighting both performance incentives and adoption pressure.

Output Metrics That Reward Speed Over Compliance

The metrics most organizations use to measure individual and team performance were designed before enterprise AI governance was a consideration, and most have not been updated to account for governance compliance as a performance variable. When a contract team is measured on contracts reviewed per week, a customer service team on tickets closed per hour, and an analyst team on reports produced per quarter, the implicit message is that speed of output is the primary performance indicator. Governance compliance that slows output reduces the metric and risks the performance evaluation.

Adding governance compliance as a performance variable is not simply a matter of adding a compliance checkbox to performance reviews. It requires restructuring the metric calculation to make compliance-adjusted throughput the relevant unit, distinguishing between work completed in compliance with governance standards and work completed outside those standards. Organizations that have done this successfully typically pair the metric adjustment with a governance enablement investment. If employees are expected to produce governance-compliant output, the governance framework must make compliant output achievable at a speed that does not fundamentally undermine productivity.

The performance metric adjustment is also an opportunity for leadership communication. When senior leaders redefine what high performance means to include governance compliance, they signal that the governance framework is a legitimate organizational priority rather than an administrative burden. That signal is more effective than policy documents at shaping the informal norms that govern day-to-day behavior.

Manager Complicity: When Team Leads Know and Look Away

One of the most common and most consequential shadow AI dynamics is manager complicity, the pattern in which team leaders know or strongly suspect that their team members are using unsanctioned AI tools and choose not to investigate, report, or intervene. Manager complicity is not usually the product of disregard for policy; it is typically the product of incentive structures that reward the team's output while insulating the manager from the governance consequences of how that output was produced.

The complicit manager faces a structurally difficult calculus: reporting the behavior would disrupt a productive team practice, create friction with high performers who are delivering results, and potentially expose the manager to questions about why the behavior was not caught earlier. Not reporting the behavior maintains team performance, avoids friction, and defers the risk. For many managers in environments without clear governance accountability, deferral is the rational short-term choice.

Addressing manager complicity requires governance designs that shift this calculus. Making managers explicitly accountable for governance compliance within their teams, not just for the team's output, changes the risk calculation. When the cost of not reporting shadow AI behavior accrues to the manager, rather than to a distant governance function, the incentive structure begins to align with the governance objective. This requires that accountability be genuine, consistently applied, and accompanied by the governance resources managers need to actually discharge it, which means fast, formal pathways for tool requests, clear guidance on what requires escalation, and training on how to have governance conversations with team members.

Diagram 2.3: Productivity Pressure and Shadow AI Adoption Loop

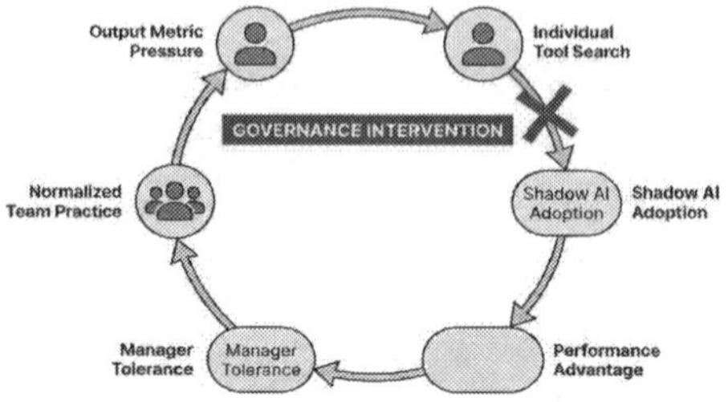

Unchecked productivity pressure risks normalizing unauthorized AI usage across teams, proactive governance interventions to break the cycle.

The Performance Review Trap

The performance review trap is a specific variant of manager complicity that occurs at the formal evaluation moment. When a manager is preparing a performance review for an employee who has been a high producer, and when that high production was facilitated in part by unsanctioned AI use that the manager informally tolerated, the review creates an acute conflict. Documenting the AI-enabled productivity accurately would require acknowledging the governance non-compliance; documenting it inaccurately to avoid that acknowledgment creates a different set of problems.

The performance review trap has organizational consequences beyond the individual review. When performance reviews consistently reward output regardless of how it was produced, the implicit message is that governance compliance is not a performance criterion. That message shapes behavior throughout the organization, not just within the team being reviewed, but also within peer teams and adjacent functions that observe the same implicit reward structure. Governance programs that seek to shift organizational behavior must work with HR to redesign performance evaluation frameworks that make governance compliance a visible, weighted component of performance assessment.

2.5 Constraint Environments and Workaround Culture

The relationship between formal constraints and workaround behavior is counterintuitive yet well-documented: the most heavily regulated environments tend to produce the most sophisticated workarounds. Public sector organizations, healthcare systems, financial institutions operating under strict regulatory frameworks, and defense contractors subject to classified information requirements all exhibit high rates of workaround behavior, precisely because the gap between what formal compliance allows and what operational efficiency demands is widest in those environments.

This dynamic is instructive for enterprise AI governance because it demonstrates that stricter controls, without enabling investments, tend to increase rather than decrease the sophistication of workarounds. Employees in high-constraint environments do not simply accept capability limitations as permanent; they develop more deliberate, technically sophisticated, and carefully constructed workarounds to avoid detection than the casual workarounds found in lower-constraint environments. Understanding this dynamic is essential for governance programs operating in highly regulated sectors.

Public Sector Productivity Constraints

Federal agencies and state government organizations face a specific combination of constraints that make AI governance particularly challenging: formal data classification frameworks that restrict what data can be processed where, procurement regulations that extend acquisition timelines, budget cycles that limit technology investment flexibility, and workforce cultures that have adapted to working within those constraints for decades. When consumer AI tools became widely available, federal employees experienced the same productivity gains as private-sector employees, but with fewer sanctioned pathways and stronger formal deterrents against unsanctioned use.

The federal government's response to this dynamic has been instructive. Multiple agencies have developed AI-use policies that attempt to provide structured pathways for employee AI use while maintaining data classification compliance. The FedRAMP framework for cloud service authorization provides a formal mechanism for evaluating AI tools against federal security standards. Still, the evaluation timeline and requirements create a significant lag between the tool's availability and federal authorization. Agencies that have addressed this through supplemental guidance, interim authorities, shadow authority frameworks, and pilot programs with defined data-handling constraints have generally achieved better compliance outcomes than agencies that relied solely on prohibition.

For managers in public sector environments, the governance challenge is compounded by workforce demographic factors: a significant portion of the federal workforce entered government service before consumer AI was a practical reality and may have limited baseline AI literacy. In contrast, a smaller portion is highly AI-capable and actively seeking productivity tools. A governance program that treats the workforce as a single group will fail both groups. Differentiated approaches that provide structured AI access pathways for higher-literacy employees while investing in literacy development for those earlier in their AI learning curve produce better governance outcomes than uniform restrictions or uniform access.

Clinical Efficiency, Pressure, and AI at the Point of Care

Healthcare is the sector where the tension between productivity pressure and governance constraints is at its highest. Clinical staff physicians, nurses, advanced practice providers, pharmacists, and clinical administrators work in environments where time pressure is constant, where documentation requirements have expanded significantly over the past decade, and where the consequences of inefficiency are measured in patient outcomes rather than productivity metrics. When AI tools offer the prospect of faster documentation, faster information retrieval, and faster decision support, the pressure to adopt is not merely about personal convenience; it is connected to a professional commitment to patient care.

The workaround behaviors that emerge from clinical productivity pressure carry risks that are categorically different from workarounds in other sectors. When a clinical staff member pastes patient information into an unsanctioned AI tool to get a faster medication interaction summary, the data governance problem potential HIPAA violation, a clinical safety problem accompanies PHI exposure to a system with unreviewed retention policies: the AI output is not subject to the validation standards applied to clinical decision support tools, and a confident-sounding incorrect output could influence a clinical decision with patient safety consequences. Shadow AI in clinical environments is a dual-risk problem.

Healthcare organizations that have made progress on shadow AI governance have typically done so by investing simultaneously in governance controls and in sanctioned AI tools that address the specific workflow needs driving workaround adoption. AI-assisted documentation tools that integrate with electronic health record systems, validated clinical decision support applications, and AI-assisted coding tools provided through formal IT channels have all demonstrated the capacity to reduce shadow AI adoption when they genuinely address the workflow needs that were driving unsanctioned tool use. Enablement investment and the governance investment are not alternatives — they are complements.

Diagram 2.4: High-Constraint Environment Workaround Sophistication

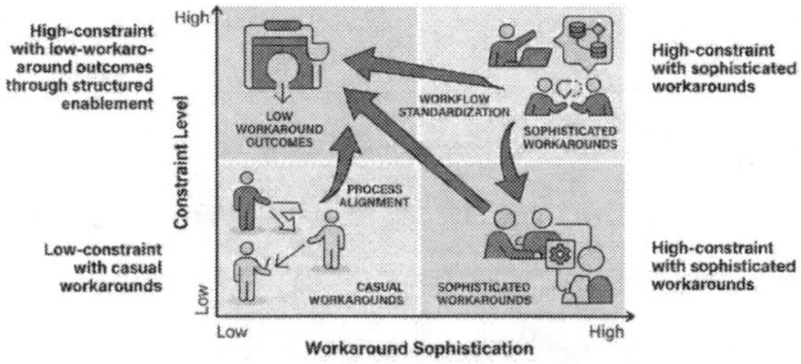

Structured governance interventions move organizations toward high-constraint environments with efficient, compliant operations and minimal workarounds.

How High-Compliance Environments Breed Sophisticated Workarounds

The sophistication gradient of workaround behavior in high-compliance environments is directly related to the governance pressure employees face. In environments where detection is likely and consequences are real, employees who choose to use unsanctioned tools develop practices designed to minimize both detection probability and consequence severity. These practices include using personal devices rather than organizational equipment, using personal accounts rather than organizational credentials, avoiding entering obviously identifiable information in favor of anonymized or decontextualized data, and conducting AI interactions during off-hours when monitoring activity is lower.

Each of these adaptive behaviors makes the governance problem harder, not easier. When employees shift unsanctioned AI use to personal devices and personal accounts, the network monitoring signals that might otherwise detect the behavior disappear. When they anonymize or decontextualize the data they enter, the individual interaction may carry a lower risk. Still, organizational visibility is lower, making it harder to understand which AI capabilities are actually being used and for what purposes. High-compliance environments that rely primarily on detection and deterrence may drive shadow AI behavior underground rather than eliminating it.

The governance implication is that organizations with high compliance obligations cannot simply import general-purpose shadow AI governance frameworks without adapting them to their specific environmental constraints. The risk calculus is different, the workaround sophistication is different, and the consequences of detection are different in ways that require tailored governance designs. The common element is the same as in lower-constraint environments: the most effective intervention is providing sanctioned alternatives that genuinely address the productivity needs driving workaround adoption.

2.6 Mapping the Legitimate Need

The most actionable analytical move in shadow AI governance is to treat unauthorized tool use as a signal of organizational capability gaps rather than as a primary compliance failure. Every instance of shadow AI use represents an employee who needed to accomplish something, evaluated the available options, and concluded that the unsanctioned option was the best available means to that end. That conclusion includes information on what the employee needed to accomplish, the approved options they considered, why those options were inadequate, and the criteria they used to select the workaround.

Extracting that information and systematically converting shadow AI behavior into data on organizational requirements requires deliberate investigative design. It is not the natural output of detection and enforcement processes, which are oriented toward documenting violations rather than understanding needs. It requires separate investigative work explicitly framed as a capability-improvement exercise rather than a compliance audit, with data protections that make employees willing to participate candidly.

Distinguishing Enablement Gaps from Policy Evasion

Not every instance of shadow AI use represents an enablement gap. Some employees use unsanctioned tools because they prefer them to approved alternatives, even when those alternatives are adequate; because they have not invested time in learning the approved tools; because they distrust organizational monitoring in general; or because they are specifically seeking to avoid oversight of particular activities. These cases represent a different problem from the enablement gap, and conflating them produces governance responses that address neither accurately.

The diagnostic question is whether a sanctioned tool that addressed the functional need the employee expressed would be adopted if provided. If the answer is yes and the employee's behavior would change with a better-approved option, the problem is an enablement gap. If the answer is no, if the employee would continue using the unsanctioned tool even with a functional approved alternative, the problem is different, likely involving preference, familiarity, monitoring avoidance, or some combination. Governance responses to enablement gaps prioritize provisioning better options; governance responses to policy evasion prioritize behavioral accountability.

In practice, most shadow AI adoption in enterprise environments falls primarily into the enablement gap category. Employees using AI tools for productivity reasons are generally responsive to sanctioned alternatives when those alternatives are genuinely capable and easily accessible. The subset of employees using AI specifically to avoid oversight is smaller and is more typically addressed through employment relations processes than through technology governance. Accurate diagnosis of the population's motivation profile determines whether the primary governance investment should be in enablement or in enforcement.

The Employee AI Survey as a Diagnostic

The most direct method for mapping the legitimate needs driving shadow AI adoption is a well-designed employee survey that fosters honest disclosure. Survey design for this purpose requires several specific features. First, it must be clearly framed as a capability improvement exercise, not a compliance audit. Employees who believe the survey data will be used to identify and discipline them will not disclose honest information about their tool use. Second, it must provide response options that do not require employees to admit to policy violations explicitly. Asking "What tasks do you wish AI tools were available for?" reveals the needed information without requiring admission.

Third, the survey should include questions about the perceived adequacy of approved tools for specific task categories, the frequency with which employees encounter tasks for which no adequate approved tool exists, and the channels through which employees currently address those tasks. Combined, these questions produce a map of functional capability gaps indexed by task type, frequency, and department. That map is the organizational requirements document that the AI tool procurement and governance design should work from.

Organizations that have conducted AI needs surveys as part of governance programs consistently report that the results surprise the governance team. Capability gaps are typically broader, more specific, and more operationally embedded than the governance team assumed, not because employees have been hiding them, but because the governance team's visibility into daily workflows is limited. The survey is an investment in that visibility, and the return is a governance program celebrating actual organizational needs rather than assumed ones.

Diagram 2.5: AI Needs Survey to Requirements Pipeline

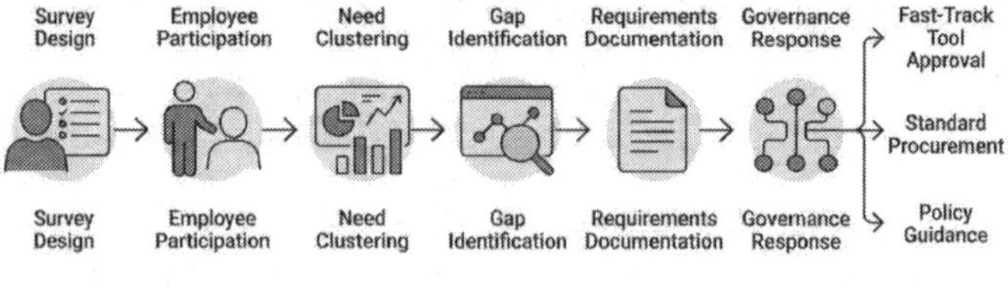

Structured approach to defining AI requirements from employee feedback.

Need Mapping: Translating Unauthorized Behavior into Requirements

Need mapping is the practice of translating observed unauthorized AI use into structured organizational requirements documents that can drive procurement, provisioning, and governance design. It requires three analytical steps: identifying the tasks for which unsanctioned AI was used, characterizing the data inputs and outputs associated with those tasks, and translating those characteristics into functional requirements that a sanctioned tool must meet.

The task identification step relies on multiple data sources: survey responses, manager conversations, technical detection data where available, and direct employee interviews in contexts where employees can speak candidly. The data input and output characterization step is analytically critical because it determines the data governance requirements that any sanctioned tool must satisfy. A task involving non-sensitive internal text has different data governance implications than a task involving patient records, financial projections, or contractual terms. The requirements document resulting from characterization specifies not just functional capabilities but also data handling, security, and compliance attributes.

The translation into requirements is the step most often missing from governance programs. Organizations that detect shadow AI use typically respond with policy communication or enforcement action; fewer invest in the analytical work of understanding what was being done and what a sanctioned pathway for it would require. That analytical investment pays governance dividends: it converts enforcement-oriented shadow AI programs into innovation-enabling governance programs that can provision tools people actually use, which is ultimately a more sustainable governance posture.

2.7 Manager's Checklist: Chapter 2

The following actions represent the critical management responsibilities for addressing productivity-driven shadow AI adoption within your team. Each item is actionable within the scope of a team-level manager, with escalation paths noted where organizational action is required.

9. Schedule a direct conversation with your team to understand what tasks currently feel inefficient and what tools, approved or not, team members are using to address that inefficiency.
10. Review your team's performance metrics to identify whether any metrics create implicit incentives for speed over governance compliance; escalate to HR or leadership if a metric redesign is needed.
11. Identify the three highest-frequency tasks for which your team currently lacks an adequate approved AI tool and document those as formal

tool requests through the appropriate procurement channel.

12. Measure the elapsed time between the last tool request your team submitted and the date the tool was available for use; use this baseline to benchmark governance responsiveness.
13. Establish explicit accountability with your team: governance compliance is a performance expectation, and workaround tool use that has not been disclosed and evaluated is not a gray area.
14. Create a mechanism for team members to raise tool requests quickly, a direct channel to you or to IT, so that the formal pathway is more accessible than the informal workaround.
15. For any team member who discloses current use of unsanctioned AI, assess whether the use involves sensitive data categories and report appropriately; treat the disclosure as constructive rather than punitive in the first instance.
16. Connect with your counterparts in HR to understand whether performance review frameworks in your organization include governance compliance as a weighted factor and, if not, advocate for that inclusion.

2.8 Key Takeaways

Shadow AI proliferates because the gap between what employees need to do their jobs effectively and what organizations provide to support that work is real, wide, and growing. The employees who bridge that gap with unsanctioned tools are not primarily acting in bad faith; they are acting in the rational interest of getting their work done, in environments where the formal pathway to better tools is slow, uncertain, and sometimes produces disappointing results.

Governance programs that treat productivity-driven shadow AI primarily as a compliance problem will invest in detection and deterrence while leaving the structural conditions that drive adoption in place. Those conditions, the enablement gap, the approval lag, the approved-tool graveyard, and the productivity metrics that reward speed without accounting for compliance, are organizational design problems that require organizational design solutions. This chapter provides the analytical framework for diagnosing those conditions accurately in your specific organizational context.

The most valuable reframe for managers and governance leaders is treating shadow AI behavior as organizational requirements data. Each instance of unsanctioned tool use contains information about what the employee needed, what approved options were inadequate, and what a sanctioned pathway would require to be competitive. Extracting that information through structured need mapping, employee surveys, and manager conversations converts a governance liability into a product design input, and the governance program that makes that conversion consistently is the one that stays ahead of shadow AI adoption rather than perpetually chasing it.

Diagram 2.6: Need Mapping to Sanctioned Pathway

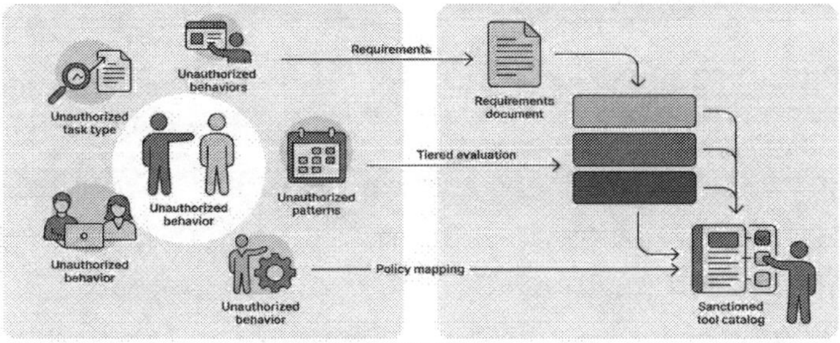

Unauthorized behavior patterns are mapped through a structured evaluation process to a catalog of approved and compliant solutions.

3 Exposure by Design

3.1 The Liability Already Accumulated

The general counsel of a regional healthcare system received an unusual request from outside legal counsel: a plaintiff's attorney had submitted a discovery request that included AI-generated communications the healthcare system did not know it had. Depositions revealed that clinical documentation specialists in three departments had used a consumer large language model to assist with drafting clinical notes for eight months. The notes had been reviewed and signed by licensed clinicians, and no clinical harm was identified. But the discovery exposed facts the organization could not easily defend: patient information had flowed into a third-party AI system the organization had never reviewed, clinical notes had been produced with AI assistance without documentation of that fact, and the organization had no AI governance policy addressing clinical documentation.

The lawsuit was ultimately settled, but the legal costs, reputational damage, and the regulatory attention the case attracted were substantial, resulting not from a harmful clinical outcome but from a governance gap. The liability had been accumulating steadily for eight months, while no one with organizational authority knew it was happening. That is the defining characteristic of shadow AI risk: it does not announce itself. It accrues quietly in the background of productive workflows until a triggering event, such as a security incident, a regulatory inquiry, a litigation discovery request, or a data breach notification, suddenly and expensively makes it visible.

This chapter maps the full risk terrain of shadow AI adoption. Understanding terrain, in specific operational terms, not as an abstract list of potential harms, but as a structured analysis of how specific exposures accumulate through specific mechanisms, is the prerequisite to building governance responses that address actual risk rather than generalized concern. The risks of shadow AI are real, serious, and quantifiable enough to support governance investment decisions. This chapter provides the analytical framework to do that.

3.2 Why This Chapter Matters

Most governance conversations about AI risk operate at an abstract level, making it difficult to translate risk awareness into action. Knowing that shadow AI creates data-leakage risk, compliance exposure, and IP concerns provides real but insufficiently specific guidance for concrete governance decisions, such as which controls to implement, which vendor terms to review, and which monitoring to set up. Operational clarity requires moving from categories to mechanisms: exactly how a specific exposure accumulates, which data inputs generate it, and which governance interventions interrupt it.

This chapter provides operational specificity across four risk domains: data leakage through AI prompts, regulatory exposure under applicable compliance regimes, intellectual property loss, and failures of decision accountability. For each domain, the analysis follows the same structure: the mechanism by which the exposure accumulates, the organizational conditions that amplify it, the sectors where it is most consequential, and the governance controls that specifically address it. The result is a risk map that managers can use to prioritize governance investments based on their organization's specific exposure profile.

Understanding your organization's risk profile is also a prerequisite for translating shadow AI governance into board-level and executive-level conversations. Risk quantification, translating technical exposure into financial, reputational, and regulatory consequence terms, is the language in which governance investments are authorized. The final section of this chapter addresses the translation challenge directly, providing a framework for converting shadow AI risk assessments into the board-ready risk communication format that governance investments require.

Diagram 3.1: Shadow AI Risk Domain Map

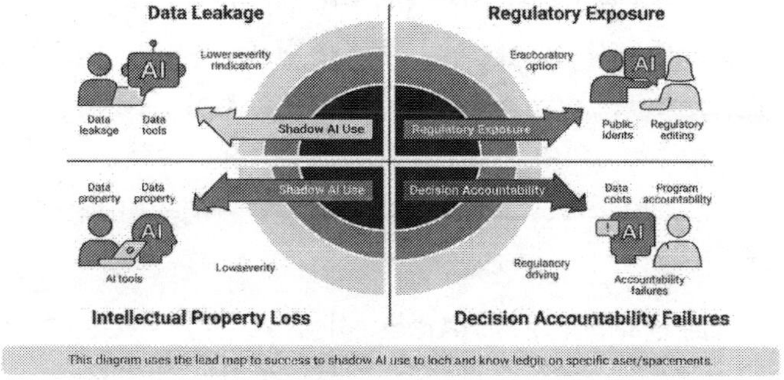

3.3 The Prompt as a Data Transfer

The fundamental data governance challenge of large language model AI is that the primary interaction mechanism, the prompt, is also a data transfer mechanism. When an employee interacts with an AI tool, they provide input to receive a generated output. That input information leaves the employee's device, traverses a network connection, and is processed by a third-party system. The data governance question, where does that information go, how long is it retained, and what are the terms governing its use, is answered by the vendor's privacy policy and terms of service, not by the organization's data governance framework.

This architecture is fundamentally different from earlier enterprise software deployments. When an employee uses an approved enterprise system, the data they enter into that system is processed according to the terms the organization negotiated and reviewed. When an employee uses a consumer AI tool, the data they enter is processed according to terms they agreed to, often by clicking through a terms of service agreement without reading it, on their own behalf, not on behalf of the organization. The organization's data governance framework does not apply to that transaction unless the organization has evaluated the tool and established a governance relationship with the vendor.

The operational implication is that every prompt an employee sends to an unsanctioned AI tool is a potential unreviewed data transfer. The word "potential" reflects that not all prompts contain sensitive data; many contain only the employee's own text, publicly available information, or task descriptions, without embedded organizational data. But the category of sensitive data that does flow through prompts is broad, and the volume of that data across a large enterprise workforce using multiple AI tools is substantial. Mapping the prompt-as-data-transfer risk requires understanding what employees actually paste into AI tools, not just what they should paste.

What Employees Paste Into AI Tools

Research on enterprise AI tool use consistently shows that employees paste a much wider variety of sensitive organizational data into AI prompts than governance teams typically assume. Common categories include draft contract language and contract excerpts, meeting notes and decision summaries, personnel feedback and performance evaluations, customer communication drafts and customer-specific context, financial projections and budget data, product roadmaps and competitive strategy documents, medical record excerpts and clinical case descriptions, legal analysis and regulatory correspondence, and code containing proprietary algorithms or system credentials.

The breadth of this list reflects a rational pattern of employee behavior. When you have access to a tool that processes natural language and produces useful outputs, you naturally bring to it the real-world context of your actual work. That real-world context is organizational data. An employee who asks an AI tool to "review this contract section for ambiguous language" and pastes a real contract section is not making a deliberate data governance decision; they are using a tool in the most natural way available. Governance that relies on employee judgment to filter sensitive data at the prompt level is relying on a control that employees are not equipped to apply consistently.

The governance implication is that prompt-level data filtering cannot be the primary control mechanism for shadow AI data governance. It requires technical controls at the tool-selection level, preventing or limiting access to tools with inadequate data-handling terms rather than at the prompt-content level. Employee training on prompt data hygiene is a useful supplementary control, but it should be understood as a supplement to technical controls, not a substitute for them. The practical guidance on prompt training is addressed in Chapter 8; the point is that the data-input reality of AI tool use requires governance responses at the system level, not the individual prompt level.

Training Data Clauses and Retention Policies

The specific vendor terms most consequential for enterprise data governance are those governing the use of training data and data retention. Training data clauses determine whether user inputs, including prompts containing organizational data, can be used to train or fine-tune AI models. Retention policies determine how long the vendor retains user input data and under what conditions it is deleted. Together, these terms define the trajectory of the organizational data from the moment it enters the prompt to the moment, if ever, it becomes inaccessible to the vendor.

Consumer AI tools vary significantly in their training data and retention terms. Some consumer AI vendors explicitly exclude user inputs from model training by default, require opt-in to use training data, and retain conversation data only for a defined period. Others retain input data for longer periods, may use it for model improvement purposes with opt-out rather than opt-in configurations, and may share aggregated or anonymized data with third parties for research purposes. For enterprise organizations with regulatory obligations governing specific data categories, these variations can be the difference between a technically compliant tool use and a reportable data incident.

Evaluating training data clauses and retention policies requires legal expertise in the specific regulatory frameworks applicable to the organization's data categories. The HIPAA prohibition on disclosing protected health information to third parties without a business associate agreement applies whether the disclosure is to a traditional software vendor or to an AI tool vendor. The contractual confidentiality obligations that protect trade secrets apply when those secrets are entered into AI prompts. Organizations that have not reviewed the training data and retention terms for the AI tools their employees use have not discharged the governance obligations required by those regulatory frameworks and contractual commitments.

Sensitive Data Taxonomies for AI Risk

Effective shadow AI data risk assessment requires a data taxonomy and a classification framework that identifies which categories of organizational data carry the highest governance implications when processed through external AI systems. The taxonomy should be structured around regulatory, contractual, and competitive sensitivity rather than general data sensitivity categories, because the governance response to each sensitivity type differs.

Regulatory-sensitive data includes categories governed by specific legal frameworks: protected health information under HIPAA, personally identifiable information under GDPR or CCPA, controlled unclassified information under federal information handling requirements, and financial data subject to SEC or FINRA disclosure rules. These categories require that any external processing occur through systems operating under terms that satisfy the applicable regulatory framework, a standard that most consumer AI tools do not meet without specific enterprise agreements.

Contractually sensitive data includes information that organizational contracts require to keep confidential: client data subject to non-disclosure agreements, supplier information covered by confidentiality provisions, and merger and acquisition information governed by deal confidentiality terms. Whether entering this data into an AI tool constitutes a breach requires legal analysis of the specific contract. Competitively sensitive data, trade secrets, product roadmaps, pricing strategies, and customer pipeline information carry reputational and financial risk beyond any contractual implications.

Diagram 3.2: Sensitive Data Taxonomy for AI Risk Assessment

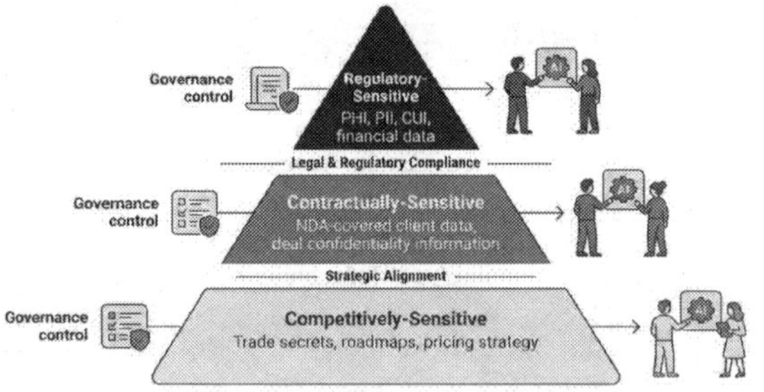

Summary the sensitive data is in notical and rent goad to lentify and documentuality assessment

3.4 Regulatory Exposure Across Compliance Regimes

Shadow AI creates regulatory exposure across multiple compliance frameworks simultaneously in most enterprise environments. A healthcare organization operating under HIPAA, employing staff who handle both federal and state government contracts, and collecting customer information governed by state privacy laws may find that a single unsanctioned AI workflow implicates three or four regulatory frameworks at once. The cumulative regulatory exposure from shadow AI adoption at scale, across hundreds or thousands of employees and multiple departments, over months of operation, can be substantial, even if each interaction appears minor in isolation.

Regulatory exposure from shadow AI differs from traditional compliance violations in one critical respect: it is often invisible until a triggering event creates disclosure obligations or investigative attention. A HIPAA breach involving physical records is typically apparent at the time of the breach. A HIPAA breach involving protected health information processed through an unsanctioned AI tool may not be apparent until a regulatory inquiry, a vendor data incident, or an internal audit specifically designed to detect AI data flows reveals the exposure. By that point, the breach reporting window may have passed, compounding the initial exposure with a disclosure timing violation.

HIPAA, PHI, and Clinical AI Workflows

HIPAA's framework for protected health information defines the conditions under which PHI can be disclosed to third parties and requires that those third parties operate as business associates and have signed a business associate agreement committing to handle PHI in accordance with HIPAA requirements. Consumer AI tools do not, by default, operate as business associates. They are designed for individual users, not for covered entities with HIPAA obligations. When a healthcare employee enters PHI into a consumer AI tool, that entry constitutes a disclosure to a non-business-associate third party and a HIPAA violation, regardless of whether the PHI is subsequently mishandled.

The clinical AI workflow dimension of HIPAA compliance is particularly complex because the clinical workflows where AI adoption is most attractive, documentation, coding, clinical decision support, and patient communication, are precisely the workflows most likely to involve PHI. A medical scribe who uses an AI tool to convert clinical notes into structured documentation, a billing specialist who uses an AI tool to identify appropriate billing codes from a clinical summary, and a care coordinator who uses an AI tool to draft patient follow-up communications are all likely to be processing PHI in their AI interactions. The productivity benefit of AI assistance in these workflows is real; so is the HIPAA exposure.

Healthcare organizations that have navigated the HIPAA-AI intersection successfully have done so by establishing enterprise agreements with vendors that operate as business associates and commit to handling PHI in accordance with HIPAA requirements. Several major AI vendors now offer healthcare-specific enterprise agreements that include business associate terms. Establishing those agreements is the minimum governance threshold; shadow AI that bypasses this structure creates the exact liability those agreements are designed to prevent.

FedRAMP, FISMA, and Federal Data Sovereignty

Federal agencies and organizations handling federal data operate under security authorization requirements designed to ensure that federal data is processed only in assessed, standards-compliant systems. FedRAMP provides the authorization framework for cloud services processing federal data; FISMA establishes the broader security framework within which FedRAMP operates. Consumer AI tools are almost universally not FedRAMP authorized at the moderate or high baselines applicable to most federal data, meaning they cannot legally process agency data, regardless of their actual security posture.

For federal agencies, the shadow AI data governance problem is not primarily a matter of vendor trust but of legal authorization. The question is not whether a particular consumer AI tool is reasonably secure; the question is whether it has the security authorization required for federal data processing. An agency employee who uses a consumer AI tool to process agency data, even data that appears non-sensitive by general standards, may violate FISMA requirements and agency security policies, regardless of the vendor's actual security posture. The legal framework does not make exceptions for good-faith productivity decisions.

Contractors handling federal data under government contracts face a related exposure. The Defense Federal Acquisition Regulation Supplement, various agency-specific data-handling requirements, and the Controlled Unclassified Information framework all impose obligations on contractors that extend to their use of AI tools. A defense contractor whose employees use consumer AI tools to assist with contract deliverables, proposal writing, or program management may be violating CUI handling requirements even when no classified information is involved. The contractor's liability extends to the prime contractor relationship, with potential consequences including contract termination for cause and exclusion from future government contracting.

Cross-Border Data Flows and Vendor Jurisdiction

AI vendors serving global markets may process user data in jurisdictions outside the user's country, subject to different legal frameworks governing government access to that data, different data protection standards, and different requirements for data deletion and subject rights. For organizations operating in the European Union, the transfer of personal data to AI vendors in jurisdictions without adequate data protection standards raises GDPR compliance questions. For organizations in sectors with data sovereignty requirements, defense, intelligence, and critical infrastructure, AI vendor processing in foreign jurisdictions may implicate national security frameworks.

The vendor jurisdiction question is not always immediately apparent from a consumer AI tool's marketing material. Data processing locations may be disclosed in privacy policies, infrastructure documentation, or data processing addenda documents that most employees never review and that most procurement processes examine only when specifically prompted to do so. Organizations with cross-border data flow obligations should include the vendor data processing location in their AI tool evaluation criteria, and employees using consumer AI tools without organizational oversight have no mechanism to verify that this criterion is satisfied.

GDPR's requirements for data processing agreements and international data transfer mechanisms, standard contractual clauses, binding corporate rules, or adequacy decisions apply to the use of AI tools when personal data is processed. An EU employee who enters personal data about a client, a colleague, or a customer into a consumer AI tool may be transferring that data to a third country without a valid transfer mechanism, creating GDPR exposure for the organization. The regulatory risk is particularly acute in HR and customer-facing functions, where personal data handling is routine and where AI productivity tools offer the most obvious workflow benefits.

Diagram 3.3: Regulatory Framework Exposure by Data Category

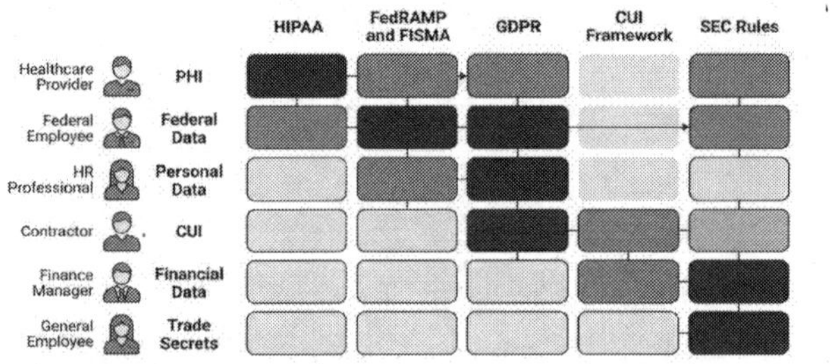

Visualize compliance risk intensity across different data types and regulations.

3.5 Intellectual Property at Risk

Intellectual property represents a category of organizational assets that is particularly vulnerable to shadow AI exposure because it is not always visibly marked as such and because the most valuable IP proprietary processes, strategic insights, competitive analysis, and product innovation are precisely the type of high-complexity, high-value content for which employees most naturally seek AI assistance. The software engineer who asks an AI tool to review a proprietary algorithm, the product manager who asks an AI to analyze a competitive positioning document, and the business development professional who asks an AI to refine a partner proposal are all creating potential IP exposure in the course of routine, high-value work.

The IP risk from shadow AI operates through two distinct mechanisms. The disclosure mechanism occurs when proprietary information enters a third-party system, potentially becomes part of a training dataset, and influences model outputs in ways that could benefit competitors. The contractual mechanism arises because many contracts governing organizational relationships contain confidentiality provisions that prohibit disclosure of covered information to third parties, and AI tools are third parties. A disclosure that breaches a confidentiality provision creates liability regardless of whether competitive harm results.

Trade Secrets and the Unintended Training Set

Trade secret protection under the Defend Trade Secrets Act and state trade secret frameworks requires that the trade secret holder take reasonable measures to keep the information secret. The "reasonable measures" requirement is a legal standard that courts apply to the specific facts of a trade secret misappropriation case. When a court considers whether an organization took reasonable measures to protect a trade secret that was subsequently disclosed through an AI prompt, the organization's AI governance posture, including whether it had a policy addressing AI tool use, whether it had trained employees on data handling, and whether it had implemented controls to prevent unauthorized disclosure, will be directly relevant to the analysis.

Organizations that have no AI governance policy, have conducted no employee training on AI data handling, and have implemented no technical controls to prevent unsanctioned AI use may find that their reasonable measures defense is significantly weakened in a trade secret case where AI exposure is a factor. The governance investment in shadow AI, the policies, training, and controls that this book describes, is not merely a risk reduction measure; it is a component of the affirmative reasonable measures defense that trade secret protection requires. Viewed through this lens, AI governance has a direct legal value that translates into risk quantification terms.

The unintended training-set concern is partially addressed by enterprise AI agreements that explicitly exclude user inputs from the training data. Several major AI vendors offer enterprise agreements with training data exclusion as a standard term, and organizations with significant IP concerns should treat training data exclusion as a minimum requirement for any AI vendor relationship. The concern is not fully eliminated by training data exclusion clauses, which address prospective training but do not eliminate retention during active use. Still, they represent a significant reduction in IP exposure relative to consumer tool terms that permit the use of training data.

Contract Confidentiality and AI Disclosure

The confidentiality provisions in commercial contracts, non-disclosure agreements, professional services agreements, client service agreements, supplier agreements, and licensing contracts were not drafted with AI tool use in mind. They were drafted to prevent the disclosure of covered information from human-to-human channels to unauthorized parties. Whether those provisions cover AI-tool disclosure is a legal interpretation question that most organizations have not formally analyzed in their specific contract portfolio, even though their employees make practical decisions about AI tool use that implicate those contracts daily.

The conservative legal interpretation of most confidentiality provisions is that entering covered information into a third-party AI tool is a disclosure to a third party and, therefore, a potential breach of the confidentiality provision. That interpretation is not universally accepted, and there are reasonable arguments for narrower readings in specific cases. But the legal uncertainty itself is a governance reason. When the breach risk is unclear, the governance posture should default toward caution, particularly for high-value contractual relationships where a breach claim would carry significant consequences.

Practical governance for contract confidentiality risk requires identifying the provisions most commonly implicated by employee AI use, client service agreements, NDA portfolios, and supplier agreements, and establishing specific guidance for employees working on covered matters. That guidance need not prohibit AI use entirely; it may specify that approved tools under evaluated vendor agreements are required for covered work, or that specific confidentiality-covered elements must not enter any AI system without express authorization. Specificity matters: broad prohibitions are routinely ignored while specific behavioral guidance is actionable.

Diagram 3.4: IP Exposure Mechanisms in Shadow AI

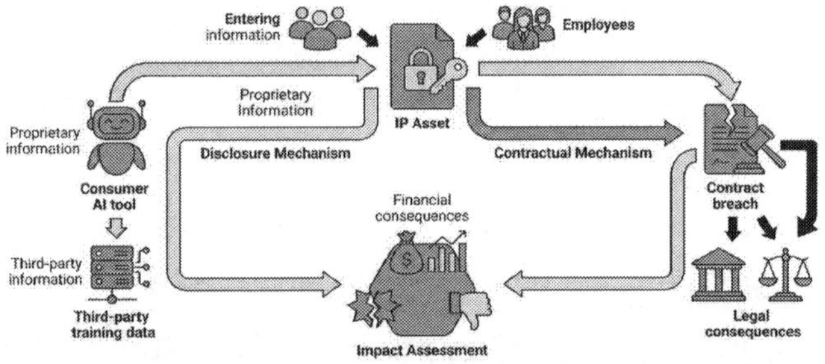

Unauthorized adoption of generative AI tools introduces operational risks through accidental data sharing and contractual violations.

Competitive Intelligence Leakage

Competitive intelligence leakage constitutes a shadow AI risk category that is difficult to detect, quantify, and attribute in retrospect. When an employee uses an unsanctioned AI tool to analyze competitive positioning, refine a competitive strategy document, or prepare for a competitive sales situation, the proprietary competitive intelligence embedded in that analysis is transmitted to a third-party system. Whether and how that intelligence might subsequently influence the AI system's outputs to other users, including employees of competitors using the same system, is a question that current AI transparency standards cannot definitively answer.

The competitive intelligence leakage risk is particularly consequential in industries where competitive information has high economic value and a short useful life, such as technology, pharmaceuticals, financial services, and defense contracting. A consumer AI tool that simultaneously processes competitive intelligence from multiple organizations is theoretically positioned at the point where competitive boundaries are algorithmically blurred. Governance does not require resolving the technical debate about the effects of training data; the appropriate posture is to treat competitive intelligence as a category that requires unconditional use of approved tools.

3.6 Decisions Without Witnesses

The data leakage risks documented above are serious. Still, they share a common characteristic: they involve something leaving the organization, data flowing outward to a system the organization does not control. The decision-accountability failures that shadow AI create are different in kind: they involve actions taken within the organization, decisions made, outputs produced, and positions taken without the documentation, attribution, and oversight that organizational accountability requires. These failures are often invisible in real time and only become visible when accountability is demanded retrospectively.

Decision accountability is a governance concept that most managers associate with financial reporting or regulatory compliance, but that has broader application in organizations where consequential decisions are made regularly. When AI-generated outputs inform those decisions, drafting communications, summarizing options, recommending actions, and producing analysis, the question of how those outputs were generated becomes relevant to accountability assessments. If the AI system generating the output was unsanctioned, unreviewed, and undocumented, the accountability chain is broken at the point of AI input.

When AI Output Becomes Organizational Position

AI-generated text that is lightly edited and issued under organizational authority, a policy document, a client proposal, a regulatory filing, or an employee communication, has become an organizational position regardless of how it was generated. The organization is accountable for the content of that position regardless of whether it was originated by a human, assisted by a sanctioned AI tool with appropriate documentation, or drafted by an unsanctioned consumer tool that no governance process has reviewed. The accountability is the same; the governance posture is dramatically different.

The problem is not that AI assistance in drafting organizational communications is inappropriate; it is not, when managed through governance structures that include review, attribution, and documentation. The problem is that unsanctioned AI use creates situations in which the organization is accountable for positions produced by processes it has not reviewed and cannot reconstruct. If a regulatory filing contains an inaccurate representation contributed by an AI-generated draft, the organization's ability to explain and defend its compliance process is compromised by the lack of documentation on how the filing was prepared.

This risk is particularly acute in regulated industries where the accuracy and good faith of organizational representations to regulators are directly assessed. A bank's regulatory filings, a pharmaceutical company's clinical trial documentation, a defense contractor's cost accounting reports, and a public company's SEC disclosures are all subject to legal standards that require the organization to be able to represent how they were prepared. Shadow AI in the workflows that generate those documents creates a documentation gap that may be difficult to close afterward.

Audit Trail Failures

Enterprise governance in most regulated environments requires that decision-relevant activities leave an audit trail, a documented record of what was done, by whom, with what information inputs, and with what authorization. AI tool use in the production of audit-trail-relevant outputs creates an audit-trail gap when the tool is unsanctioned and undocumented. The gap is not merely technical; it is not simply that the AI tool's log files are unavailable. It is structural: there is no governance mechanism to document, review, or validate the AI tool's role in producing the output.

Audit-trail failures in shadow AI contexts span organizational functions. In financial services, trading decisions informed by AI-generated analysis lack complete audit documentation. In legal functions, AI-assisted contract reviews go unattributed. In human resources, AI-assisted hiring decisions cannot be reconstructed with the level of detail required by equal employment opportunity compliance. In each case, the underlying activity is not inherently inappropriate; the governance failure is the absence of a documented, attributable process.

Building audit trail governance for AI use requires that any AI tool operating in audit-trail-relevant workflows be sanctioned through a process that includes review of the tool's logging and documentation capabilities. Tools that cannot produce reviewable logs of their interactions, retain logs only for short periods, or do not provide organizational administrators with log access cannot satisfy audit trail requirements for regulated workflows. This represents a specific procurement criterion that AI governance programs in regulated industries should apply consistently.

Liability When AI Is Neither Approved nor Documented

Courts and regulators are still resolving the intersection of shadow AI and organizational liability, but several patterns are already visible. Where AI-assisted outputs inform decisions that are subsequently challenged, employment decisions, clinical decisions, financial decisions, and regulatory submissions are the most common categories. The organization's liability exposure is shaped by what it knew or should have known about how the decisions were made. Organizations that lack an AI governance policy and monitoring capabilities may face a "should have known" liability standard in sectors where AI governance has become an expected baseline.

The regulatory dimension is evolving. The EU AI Act establishes risk-based requirements for AI used in high-risk applications, and similar frameworks are being developed elsewhere. Organizations using AI in consequential decisions, such as hiring, credit, healthcare, and law enforcement, are expected to document that AI systems meet defined governance standards. Shadow AI in high-risk workflows creates a categorical compliance failure: not just a failure to meet the standards, but a failure to have documentation that demonstrates whether the standards are met at all.

Diagram 3.5: Decision Accountability Gap in Shadow AI Workflows

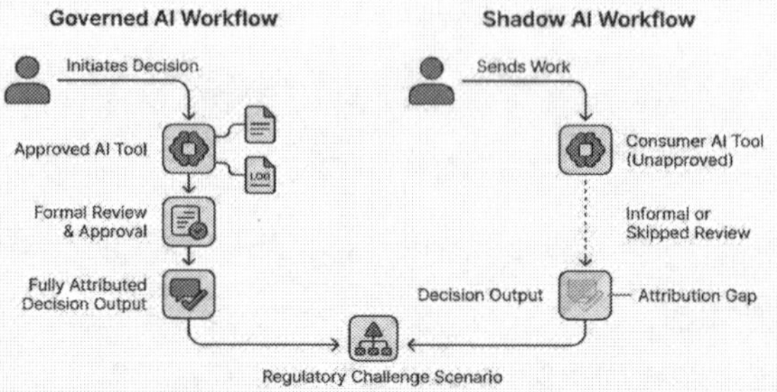

Lack of visibility and controls in Shadow AI creates accountability gaps, complicating responses to regulatory challenges.

3.7 Quantifying What You Cannot Easily Measure

One structural challenge of shadow AI governance is that its risks are largely latent; they have not yet materialized as quantifiable losses, making it difficult to justify governance investments through the financial metrics that budget decisions require. The analytical problem is how to convert uncertain future risk into a governance investment case that decision-makers can evaluate with the same rigor applied to capital expenditures. This challenge is not unique to shadow AI. Still, it is particularly acute here because the triggering events that make latent risk visible are also the events governance is designed to prevent.

Shadow AI risk quantification is achievable to a degree of precision sufficient for governance investment decisions, even though it cannot achieve the actuarial precision available for well-documented risk categories. The quantification approach uses expected value analysis, multiplying the probability of a risk event by its estimated financial consequence, across the specific risk categories relevant to the organization's regulatory environment, data sensitivity profile, and AI adoption pattern. The result is a range of expected annual risk exposure that can be compared with the cost of governance controls to evaluate the efficiency of governance investment.

Shadow AI Risk Scoring Framework

A practical shadow AI risk scoring framework evaluates three variables for each risk category: exposure probability, which reflects how likely it is that the risk event will occur given current AI adoption patterns; consequence magnitude, which reflects the financial, regulatory, and reputational cost of the risk event if it occurs; and control effectiveness, which reflects the degree to which existing governance controls reduce the probability or magnitude of the risk event. The product of exposure probability and consequence magnitude, adjusted for control effectiveness, produces an expected risk value for each category.

Exposure probability assessment draws on three data inputs: estimated prevalence of shadow AI use, assessable through network traffic analysis, employee surveys, and vendor access logs; the sensitivity profile of data flowing through those tools, assessed through data classification exercises; and vendor terms governing that data, requiring review of privacy policies for the tools identified in the prevalence assessment. Organizations that skip prevalence assessment are estimating probability without critical inputs, which systematically produces underestimates.

Consequence magnitude assessment requires sector-specific knowledge of applicable penalty structures. HIPAA breach penalties follow a tiered civil money penalty framework. GDPR fines are calculated as a percentage of global annual revenue. FedRAMP authorization violations can result in contract termination and debarment from future government contracting. Trade secret misappropriation claims carry compensatory and exemplary damages. Quantifying the magnitude of consequences requires legal expertise in the applicable frameworks and a defensible estimate of the organizational exposure shadowed by AI.

Diagram 3.6: Shadow AI Risk Scoring Matrix

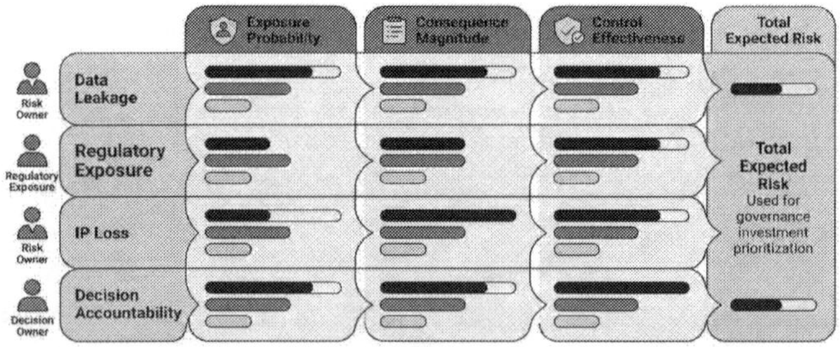

Diagram 3.6: Shadow AI Risk Scoring Matrix scoring scores used for governance investment prioritization.

Translating Technical Exposure into Board-Level Risk

The translation from technical risk assessment to board-level risk communication is a governance skill that requires deliberate development. Board members are typically not equipped to evaluate technical shadow AI risk in technical terms; they evaluate risk in terms of financial exposure, reputational consequence, regulatory liability, and competitive disadvantage. The governance professional's task is to translate the technical assessment into those terms without losing the specificity that makes the assessment actionable.

The most effective board-level shadow AI risk communications share three characteristics. First, they anchor the risk in a specific organizational context, naming functions, data categories, and estimated regulatory exposure under applicable frameworks. Second, they benchmark the organization's posture against peer organizations or regulatory requirements, providing the board with a reference point for evaluation. Third, they include a governance investment proposal with estimated risk reduction outcomes, so the board evaluates the investment in terms of risk purchased rather than cost incurred.

Board-level risk communication for shadow AI is also an opportunity to establish governance as a strategic function rather than a compliance function. Organizations that frame AI governance as a competitive enablement investment, the governance infrastructure that allows the organization to capture AI productivity responsibly, tend to receive stronger executive and board support than organizations that frame governance as risk mitigation alone. Both frames are accurate; the enablement frame activates a different organizational response.

3.8 Manager's Checklist: Chapter 3

The following actions represent the specific risk-management responsibilities of team-level managers within the shadow AI governance context. Each item addresses a specific exposure mechanism described in this chapter.

17. Identify which regulatory frameworks apply to the data your team regularly handles: HIPAA, GDPR, FedRAMP, CCPA, or sector-specific frameworks, and assess whether those frameworks impose requirements on AI tool use.
18. Review the data sensitivity profile of your team's most frequent AI-use tasks: what categories of data are being entered into AI prompts, and do those categories require specific governance treatment under applicable regulatory or contractual frameworks?
19. Identify any contracts your team supports, client agreements, supplier contracts, NDAs, or government contracts that contain confidentiality provisions, and assess whether any current AI tool use might implicate those provisions.
20. Assess whether any decisions made in your team's workflows are informed by AI-generated outputs, and determine whether those decisions require audit trail documentation that current AI tool use does not provide.

21. Determine whether your team produces any outputs, such as regulatory filings, client proposals, compliance documentation, or formal organizational communications, that require attribution and process documentation, and evaluate whether current AI-assisted workflows in those areas satisfy that requirement.
22. Ensure that team members understand the three categories of data that should not enter any unsanctioned AI tool under any circumstances: regulated data governed by applicable compliance frameworks, information covered by confidentiality provisions in organizational contracts, and competitive intelligence that constitutes organizational trade secrets.
23. Report any current shadow AI use in your team that involves regulated data categories to the appropriate governance, legal, or compliance function; the disclosure is a governance improvement action, not an admission of wrongdoing, and the cost of not disclosing is higher than the cost of disclosing.
24. Advocate within your management chain for the governance investment required to provide your team with sanctioned AI tools that address the specific workflow needs driving current or potential workaround adoption.

3.9 Key Takeaways

The shadow AI risk landscape is specific and accumulating, and in most enterprises, it is already larger than the governance team knows. Risks span four domains: data leakage through the prompt mechanism, regulatory exposure under applicable compliance frameworks, intellectual property loss due to confidentiality breaches and training data exposure, and decision accountability failures that prevent organizations from reconstructing consequential decisions. Each domain has a distinct exposure mechanism, distinct governance controls, and distinct organizational conditions that amplify or constrain it.

The practical governance implication of this risk analysis is prioritization. Governance resources are finite, and not every risk category is equally consequential for every organization. The risk scoring framework identifies which categories carry the highest expected exposure given your specific regulatory environment, data profile, and AI adoption pattern. That analysis should drive governance investment: highest-priority categories receive controls and monitoring first, while lower-priority categories receive lighter attention until the primary exposures are addressed.

The most important managerial insight is that shadow AI risk is not a future concern; it has a past tense in most organizations. Liability has already begun accumulating, for whatever period and proportion of the workforce has been using unsanctioned tools. The governance work ahead is partly curative, partly interceptive, and partly preventive of further accumulation. Starting sooner rather than waiting for a triggering event to make the urgency visible is the risk-rational course.

4 The Discovery Problem

A financial services firm retained a third-party audit firm to assess its AI exposure after an internal report flagged unexpected data in a large language model vendor's breach notification. The auditors expected to find a handful of unsanctioned tools concentrated in a single department. Instead, they uncovered 47 distinct AI endpoints in active use across 11 business units, tools discovered through a combination of DNS log analysis, expense report mining, and confidential employee interviews, with an amnesty guarantee. Of those forty-seven tools, fewer than six had ever appeared in any procurement record. The organization had, in effect, been operating a shadow AI environment larger than its entire sanctioned technology stack for nearly two years, without any structured visibility into what was happening.

That audit experience captures the central challenge of AI governance at the discovery stage. Before an organization can govern its AI usage, it has to find it. And finding it is considerably harder than most governance programs assume. Shadow AI does not announce itself. It does not create formal procurement records. It does not trigger change management workflows. It enters the organization through browser extensions, SaaS platform updates, personal subscriptions that employees quietly renew on corporate credit cards, and vendor products that embed new AI capabilities into already-approved features. By the time a governance team becomes serious about inventory, it is already behind.

This chapter offers a structured approach to the discovery challenge, one that yields a reliable inventory of AI tools in operation without converting the effort into a surveillance operation that damages employee trust. Discovery is not a one-time event. It is a continuous function. But it has a starting point, and that starting point is almost always messier and more revealing than organizations expect.

4.1 The Inventory Paradox

The first obstacle to AI discovery is the expectation that existing technology management processes will surface the problem. Most organizations already have endpoint management solutions, software asset management platforms, and network monitoring tools. These systems are designed to track what is installed on managed devices and what traffic is passing through the corporate network. Managers frequently assume that these tools would flag significant AI usage if it were occurring. That assumption, it turns out, is wrong, and understanding why it is wrong is the starting point for building a more accurate discovery methodology.

The gap exists because shadow AI is predominantly web-based. Unlike legacy shadow IT, which typically involved installed software that endpoint management could detect, modern AI tools operate almost entirely through browsers. An employee who visits ChatGPT, Claude, or Perplexity does not install anything that asset management software can see. The transaction is indistinguishable from visiting any other web application. Standard endpoint management has no native mechanism to classify browser-based AI access as categorically different from accessing a news website or a productivity tool. The tool is not on the device; it is on the internet, and access to it leaves only the thinnest of traces in standard monitoring logs.

Diagram 4.1: The AI Inventory Gap

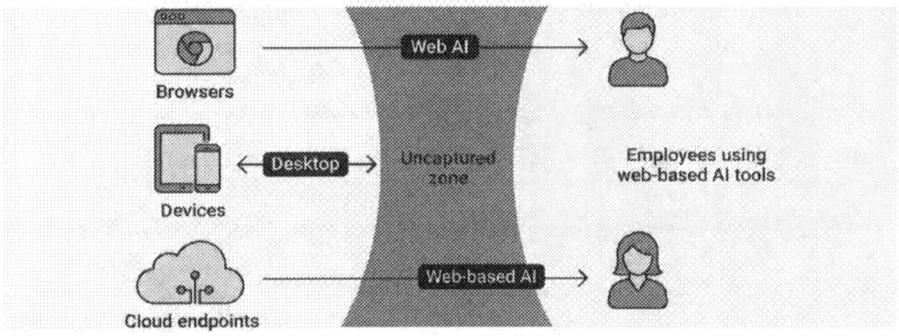

The space between what enterprise endpoint management systems can see and what is actually in use.

Web-Based AI Access and Endpoint Management Limits

Endpoint detection and response tools, software asset management platforms, and unified endpoint management solutions share a common design assumption: that software of interest runs locally on managed hardware. That assumption was reasonable when shadow IT consisted of pirated software or unauthorized applications installed on corporate laptops. It is not reasonable when the primary risk surface is a browser tab. An employee submitting a sensitive contract clause to a web-based AI tool generates no local artifact that any standard endpoint tool can inspect. The entire transaction occurs in the cloud and returns a result through the browser. The endpoint management system sees a browser process consuming CPU cycles, nothing more.

Some organizations have attempted to address this gap by managing browser extensions, requiring only approved extensions to operate on corporate devices. This helps at the margin but does not close the core problem. Browser-based AI access does not require an extension. It requires only a URL and a login. And for employees using personal devices even briefly, for work-related tasks, there is no managed endpoint at all. The transaction is entirely invisible to corporate technology infrastructure.

This is not a failure of technology management teams. It is a structural mismatch between the tools those teams inherited and the new reality of how AI is accessed. The correct response is not to blame endpoint management for failing at a task it was never designed to perform, but to supplement it with discovery methods calibrated to the actual characteristics of AI tool access.

BYOD Environments and Device-Level Blind Spots

The bring-your-own-device policy choices many organizations made during the remote-work expansion of the early 2020s created lasting visibility gaps that now define the outer boundaries of AI discovery. In organizations where employees routinely use personal smartphones, tablets, or laptops for work-adjacent tasks, the endpoint management coverage that security teams have worked to maintain on corporate hardware does not extend to those personal devices at all. A knowledge worker who begins an AI-assisted analysis on a corporate laptop, continues the conversation on a personal phone during a commute, and finalizes the output on a home desktop has generated three separate AI interactions, and the organization may have visibility into only the first.

The practical implication for discovery programs is that device-level detection alone cannot produce a reliable inventory. Any methodology that relies exclusively on what can be seen at the endpoint will systematically undercount AI usage by the magnitude of personal device activity in the organization. For organizations with large field forces, remote-first workforces, or significant contractor populations, that undercount can be substantial, potentially representing a majority of total AI interactions.

This is not an argument for expanding device management to personal hardware. In most organizations, that path is legally complex, technically invasive, and culturally corrosive. It is, instead, an argument for building discovery programs that combine technical detection on managed endpoints with non-technical methods, disclosure programs, manager-led inventory, and procurement signal analysis—to capture what technical methods cannot reach.

Shadow AI in Sanctioned Platforms

A particularly difficult discovery problem involves AI capabilities embedded in software that the organization has already approved. When a project management platform adds an AI assistant to its existing feature set, or when a customer relationship management system activates a generative summarization feature through a platform update, the resulting AI usage is categorically different from an employee who independently signs up for a separate AI service. But the governance risk may be identical to, or greater than, that of the standalone service because the embedded feature carries an implicit organizational endorsement that the standalone service lacks.

Managers and employees frequently assume that if a tool has been approved, its features are approved as well. That assumption is rarely examined systematically. SaaS vendors regularly expand their feature sets through routine updates, and those expansions increasingly include AI capabilities that were not present at the time of procurement review. An organization that conducted a thorough vendor assessment in one year and approved a platform may be running meaningfully different AI features within eighteen months, features that process the same sensitive data but under different terms, accepted through a click-through update agreement rather than a legal review.

This phenomenon of shadow AI within sanctioned platforms is, in some ways, more dangerous than standalone tool adoption because it is entirely invisible to standard detection methods. The network traffic goes to an approved domain. The access patterns are normal. The procurement record is clean. Only a deliberate review of vendor feature change logs and updated terms of service would reveal the shift. Building that review into the discovery program is not optional; it is where a substantial portion of undiscovered AI usage actually lives.

4.2 Network and Procurement Signals

While endpoint management has fundamental limitations in detecting web-based AI usage, organizations are not without technical detection options. Two sources of organizational data, network traffic logs and financial records, provide meaningful signals that, when analyzed systematically, can substantially narrow the inventory gap. Neither source alone is sufficient. But together, and combined with the human-facing methods described in the next section, they form the backbone of a credible discovery program.

The key insight is that AI tool usage leaves traces in organizational systems even when it leaves none at the endpoint level. Every browser-based AI interaction passes through the organization's network infrastructure if conducted on a corporate network or through a corporate VPN. Every subscription paid with a corporate credit card or reimbursed as a business expense creates a financial record. And every vendor contract that now includes AI features has a contract file somewhere in the organization's legal or procurement system. The discovery program's task is to surface and connect those traces.

Diagram 4.2: The Three-Signal Discovery Model

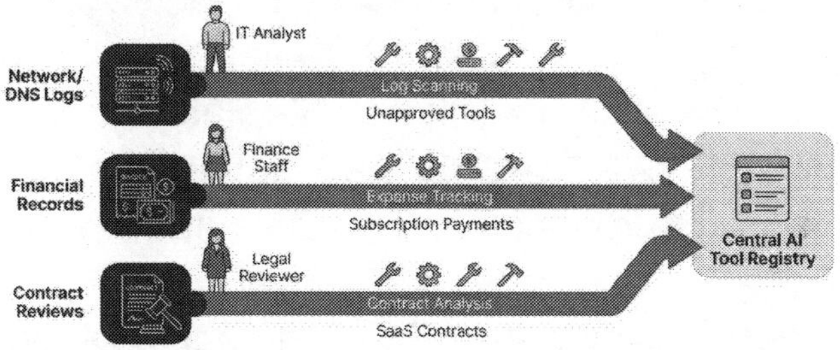

Multiple data streams are aggregated to build a comprehensive inventory of utilized AI tools across the organization.

DNS and Web Proxy Logs as Detection Instruments

Domain Name System logs and web proxy logs represent the most technically accessible route to AI discovery for organizations with managed network infrastructure. Every time a corporate device resolves a domain associated with an AI tool, openai.com, anthropic.com, gemini.google.com, and hundreds of others, that resolution event creates a record in the DNS infrastructure. When deployed, web proxy logs record the full URL of requests passing through the proxy, providing an even richer signal. Neither source requires additional software procurement; both are generated continuously by infrastructure that most medium and large organizations already operate.

The practical challenge is not data availability but classification and volume. Corporate DNS logs generate enormous quantities of entries daily, and the domains associated with AI tools are not always obviously named. Some AI capabilities are served from CDN domains or cloud platform subdomains that appear generic without context. Building and maintaining a curated list of AI-related domains and keeping it up to date as new services emerge is a recurring operational task rather than a one-time configuration. Several commercial threat intelligence feeds now include AI service domain categorization specifically for this purpose, and building that categorization into existing SIEM or network monitoring infrastructure is achievable in most organizations within weeks.

DNS analysis produces a frequency-and-breadth picture: which AI domains are being accessed, from how many distinct source devices, and at what volume. It does not reveal the content of interactions. It should not, both for privacy reasons and because capturing prompt content at the network layer is technically complex and legally sensitive in many jurisdictions. The goal is inventory and frequency data, not content surveillance. A manager reviewing DNS-derived AI usage data should look for patterns that indicate which departments are generating the most AI traffic, which tools appear most frequently, and whether any tools raise immediate compliance questions.

Expense Report Mining

Financial records provide a discovery signal that technical network analysis cannot. An employee who uses AI tools exclusively on a personal device connected to a home network will generate no corporate network traffic whatsoever. Still, if that employee seeks reimbursement for an AI subscription or uses a corporate credit card to pay for it, a financial record exists. Expense report analysis is, therefore, a critical complement to network-based detection, capturing a population of AI users that technical methods may miss entirely.

Implementing expense report analysis for AI discovery requires two things: a categorization scheme that flags AI-related expenses in accounts payable and expense management systems, and the organizational authority to request that information. Most organizations already have expense categories for software and subscriptions; refining those categories to flag AI tool subscriptions specifically is a straightforward configuration change in most expense management platforms. The harder challenge is typically organizational, establishing that the information will be reviewed and used for inventory purposes rather than for individual employee discipline, which requires the same trust architecture discussed in the employee disclosure section.

Expense report mining tends to surface different tools than DNS analysis. Network logs capture frequent, habitual AI usage conducted on corporate infrastructure. Expense records capture deliberate subscription-based usage that employees consider valuable enough to pay for and claim back. The intersection of the two sources is informative; the tools that appear in both sources have the deepest organizational footprint and therefore the highest priority for governance attention.

Vendor Contract Reviews as AI Feature Audits

The third technical discovery signal lies in the organization's own vendor contract portfolio. Most enterprise software agreements now include provisions governing AI features, data training rights, and model usage policies. Still, those provisions are frequently buried in addenda, supplemental terms, or unilateral policy documents that were accepted through click-through at the time of a platform update. A systematic review of active vendor contracts with an AI-feature lens will almost always surface capabilities the organization did not know it had purchased and risks it did not know it had accepted.

This review is best conducted as a joint exercise between legal, procurement, and security teams. Legal reads for data handling and training rights clauses. Procurement identifies which vendors have introduced AI features since the last contract review. Security evaluates the access permissions that those AI features may require to function, whether they need access to stored documents, email archives, or other sensitive data stores that feed their context windows. The output is a list of AI capabilities already active within approved platforms, distinct from but equally important as the list of standalone AI tools accessed through network channels.

Diagram 4.3: Vendor Feature Expansion Timeline

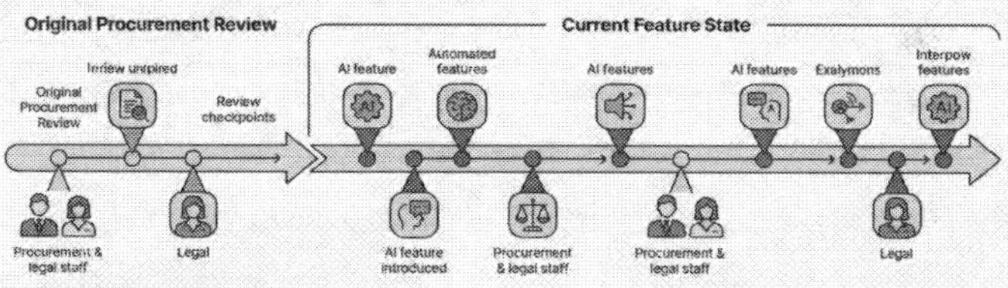

A visualization tracking the iterative deployment of AI features and regulatory review checkpoints throughout a standard two-year vendor contract agreement.

4.3 Employee Disclosure and Trust-Based Inventory

Technical discovery methods are necessary but not sufficient. They capture what leaves a trace in organizational infrastructure; they miss what occurs on personal devices, through consumer accounts, or in the cognitive work that employees perform using AI outputs, without those interactions ever passing through corporate systems. Capturing that dimension of AI usage requires a fundamentally different approach: asking employees directly, in a way that makes honest disclosure feel safe and professionally reasonable.

Trust-based inventory is the most contentious component of a shadow AI discovery program because it requires organizational leaders to make a genuine commitment that disclosure will not be used punitively. That commitment is easy to state and hard to maintain, particularly when disclosure reveals usage patterns that make compliance and security leaders uncomfortable. But the programs that have produced the most complete inventories, including the financial services audit described at the opening of this chapter, have invariably included a well-designed employee disclosure component. The technical methods find what they can find. The human methods find the rest.

Designing AI Disclosure Programs

A functional AI disclosure program has three design requirements. First, the mechanism must be simple enough that completing it takes less time than the employee believes compliance is worth. Long, multi-page surveys with detailed technical questions about tool configurations and usage frequencies will be abandoned or completed dishonestly. The initial disclosure instrument should ask for the name of any AI tools in use, the approximate frequency of use, and the general type of content being processed, and nothing more. Supplemental questions can follow for tools that surface as high-priority.

Second, the program must communicate its purpose clearly and explicitly. Employees will not disclose AI usage if they believe the information will be used to discipline them, restrict their tools before alternatives are provided, or create permanent records that follow them through performance evaluations. The communication framing must address those concerns directly, stating explicitly what the information will be used for, who will see it, and what protections exist for employees who disclose tools that may have been used in ways that technically violate existing policy. Vague assurances of confidentiality will not produce honest responses; specific, institutionally credible commitments will.

Third, the disclosure must be positioned as professionally meaningful rather than administratively burdensome. Employees who have been using AI tools productively often want to share what they have discovered. They sometimes, correctly, believe that the tools they have found represent genuine operational value. A disclosure program framed as the organization's attempt to understand what is working and what employees need creates a very different response than one framed as an audit or a compliance exercise. The framing is not window dressing; it determines the quality of the data the program produces.

Diagram 4.4: AI Disclosure Program Design

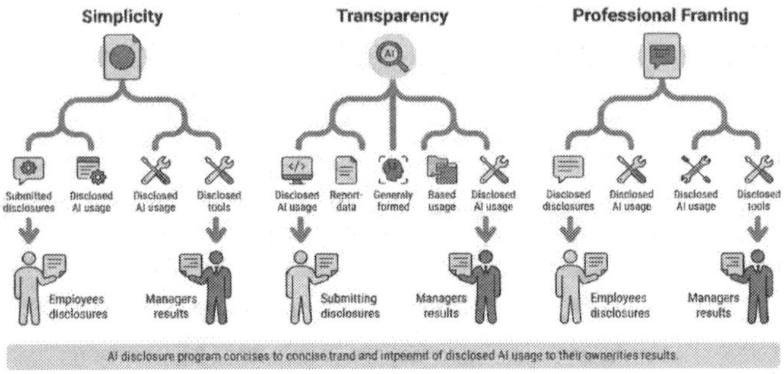

Amnesty Frameworks for Self-Reporting

The single most effective mechanism for producing honest AI usage disclosure is a formal amnesty framework, a documented, institutionally authorized commitment that self-reported usage during a defined period will not be subject to disciplinary review, regardless of whether that usage technically violated existing policy. Amnesty frameworks are modeled on practices that compliance programs in regulated industries have used for decades, such as in tax, healthcare billing, and securities reporting. The basic logic is consistent. Organizations obtain better information when reporting does not trigger punishment, and better information produces better risk management than punitive enforcement of policies that could not be enforced anyway.

A well-designed AI amnesty framework specifies the disclosure window, the scope of protection (typically covering use of any AI tool not involving intentional data theft or fraud), the mechanism for disclosure, and the organizational authority under which the amnesty is granted. The last element is critical. An informal promise from an IT manager carries little weight. A formal amnesty notice, signed by the general counsel and chief human resources officer and communicated through official channels, carries substantial weight. Employees evaluating whether to disclose are making a judgment about institutional reliability, not about individual good intentions, and the institutional signal matters accordingly.

Organizations that have deployed amnesty-backed disclosure programs consistently report inventories that are two to four times larger than those produced by technical detection alone. The additional tools surfaced by amnesty programs are disproportionately concentrated in leadership-adjacent roles, executives, and senior managers who have used AI tools with significant business impact and have the judgment to assess whether amnesty protections are credible. Capturing that population's usage is particularly valuable because those individuals tend to be processing the most sensitive categories of organizational information.

Manager-Led Team Discovery

In parallel with the formal disclosure program, manager-led team discovery provides a direct, human-mediated inventory channel that can reach employees who are skeptical of formal institutional disclosure mechanisms. In this approach, team managers conduct structured one-on-one conversations with each direct report, using a brief standardized guide to ask about AI tool usage, perceived value, and any concerns the employee has about whether their usage is appropriate. The manager is positioned not as an auditor but as a resource, someone who is themselves trying to understand what tools their team has found useful and what support they need.

Manager-led discovery has several advantages over centralized disclosure programs. The interpersonal trust between a manager and a direct report typically exceeds the trust employees extend to institutional processes. Managers can respond in real time to employee concerns, clarifying what will. They will not happen with disclosed information and by addressing the specific anxieties that prevent disclosure in formal survey formats. And the conversation format allows managers to surface contextual information about how the tool is being used, in which workflow, with what type of content, and which structured survey instruments are not captured.

The prerequisite is that managers themselves must be trained and briefed before conducting these conversations. A manager who responds to a disclosure with visible alarm or who promises consequences they are not authorized to impose will produce the opposite of the intended effect. The manager training for team discovery should cover the questions to ask, the responses to give, the promises that can and cannot be made, and the escalation path for situations that require specialist input from legal or security. The investment in that training pays dividends both in discovery quality and in the managerial relationship capital that the process either builds or depletes.

4.4 Building a Living AI Tool Registry

Discovery produces data. The data needs a home. An AI tool registry is the organizational artifact that converts discovery findings into something actionable, a maintained, searchable record of AI tools in operation that informs governance decisions, risk assessments, vendor evaluations, and policy development. Building the registry correctly from the outset avoids the technical debt that plagues many first-generation AI governance programs, where discovery data is collected but never structured in a way that makes it useful across functions.

The registry is not a spreadsheet, though it may begin as one. It is a living record designed for multiple audiences with different information needs: IT needs to know the technical access profile of each tool; legal needs to know the data handling terms; compliance needs to know which regulatory regimes each tool's usage may implicate; and business unit leaders need to know which of their teams are using which tools and whether those tools are currently in a permitted, provisional, or prohibited status. A registry architecture that serves all of those audiences requires intentional design before data entry begins.

Diagram 4.5: AI Tool Registry Architecture

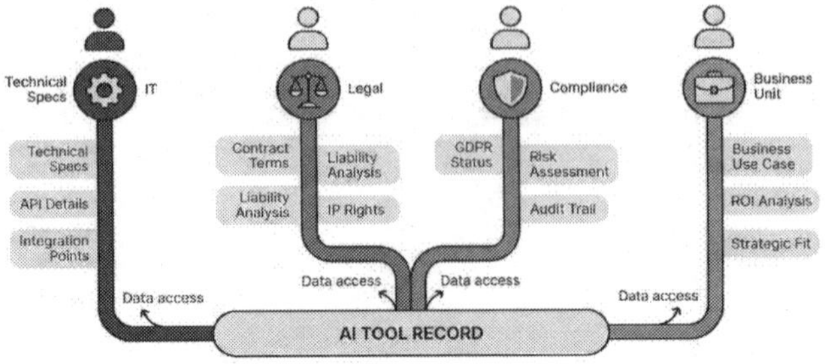

Role-based visibility ensures specific functional teams access relevant attributes for efficient AI tool governance.

Registry Architecture: What to Capture and Classify

A functional AI tool registry entry should, at a minimum, capture the following categories of information. Identification data includes the tool name, vendor, URL or access point, and primary use category: productivity, code generation, data analysis, content creation, customer communication, or similar. Access profile data includes how the tool is accessed (browser, API, embedded within another platform), whether it requires authentication, and what data the tool processes or retains. Organizational data includes which business units are using the tool, the approximate user count, and the operational point of contact.

Risk classification is a separate field that should be populated through a structured assessment rather than intuition. A simple three-tier classification, low, medium, and high, applied consistently across all registry entries, provides the sorting layer that governance teams need to prioritize their work. Low-risk tools are those that process only publicly available or non-sensitive content, with clear no-retention terms and no regulatory exposure. Medium-risk tools process internal but non-sensitive content or have ambiguous retention terms. High-risk tools process sensitive personal data, regulated information, confidential business information, or have terms that explicitly permit training on user inputs.

Status is the most operationally important field in the registry. Each tool should carry one of four status designations: permitted, provisionally permitted, under review, or prohibited. The status field is what makes the registry a decision support tool rather than a historical record. When a business unit asks whether a particular AI tool is approved for use, the registry should be the authoritative source, not a policy document that requires interpretation, nor a conversation with an IT manager who may not be up to date. The registry is the enterprise's real-time AI governance instrument.

Continuous Discovery: Keeping the Inventory Current

An AI tool registry built once and not maintained becomes inaccurate faster than almost any other type of technology inventory because the AI tool landscape changes more quickly than most enterprise software categories. New tools emerge weekly. Existing tools add features that change their risk profile. Vendor terms of service are updated. Tools that were provisionally permitted may become ineligible as their vendors change data handling practices. Keeping the registry current requires treating discovery as an operational function rather than a project.

The operational model for continuous discovery combines the three technical signals described earlier: DNS log analysis, expense report review, and vendor contract monitoring, with a quarterly manager-led check-in process and an ongoing employee disclosure channel that remains open after the initial inventory exercise. Each of those channels has a different cadence. DNS analysis should run on a monthly or weekly cycle, automated where possible. Expense reviews typically align with accounts payable processing cycles. Vendor contract monitoring should trigger whenever a vendor issues a platform update notice or terms of service update. Manager check-ins provide a human signal quarterly. The disclosure channel remains permanently open for employees who want to disclose new tool adoption without waiting for the next formal cycle.

Diagram 4.6: Continuous Discovery Operating Model

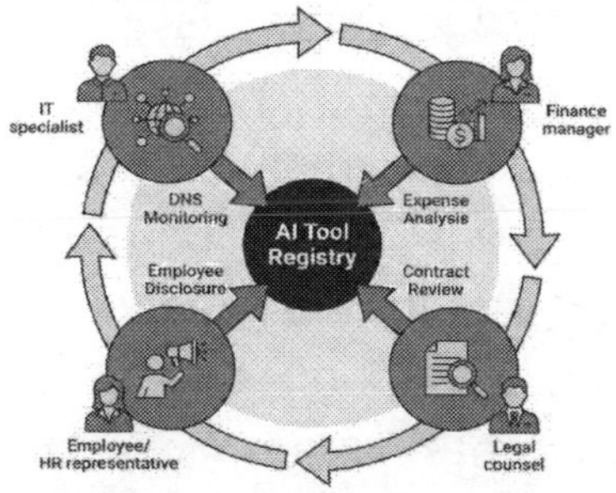

Ongoing identification of AI tool use within the organization through multiple integrated discovery sources and central tracking.

The governance question that continuous discovery must answer is not just what new tools have appeared, but what has changed about existing tools. A tool that was low-risk when originally registered may become high-risk if the vendor acquires a company with different data practices, changes its terms to permit model training on user inputs, or is acquired by a foreign entity subject to data localization requirements in a different jurisdiction. Monitoring for those changes requires the same vendor surveillance discipline that legal and procurement teams apply to contract compliance, systematic, scheduled, and assigned to named individuals with clear accountability.

Communicating Findings Without a Chilling Effect

The final and often underestimated challenge of discovery is communicating what the program found without triggering the defensive behaviors that undermine future transparency. When an organization completes its first AI inventory and discovers extensive unsanctioned usage, there is a natural institutional impulse to respond with restrictions to immediately move tools from the discovered status to prohibited, to send enforcement-tone communications, or to announce new monitoring measures. That impulse, if acted on without care, converts the discovery program into the instrument of organizational surveillance that employees feared it was, guaranteeing that future disclosure programs will produce only the information that employees calculate is safe to share.

Findings communication should separate the inventory results from the governance response. Reporting to organizational leadership on the scope and nature of shadow AI usage is a necessary step, but the governance response, which tools will be reviewed for potential approval, which will be restricted, and what alternative tools will be provided, should follow from that review rather than being announced simultaneously with the discovery findings. Employees who disclosed tools in good faith during the amnesty period deserve to know that their disclosure is being evaluated thoughtfully, not that their tools are being shut down immediately.

The communication to the workforce following discovery should acknowledge the scope of what was found, express the organization's appreciation for employees who disclosed usage during the amnesty period, and describe the next steps in the review process. It should explicitly state what amnesty commitment means for employees whose disclosed tools are ultimately placed in prohibited status, that the disclosure itself remains protected even if the tool's status changes. And it should provide a timeline for registry population and the launch of a formal tool-approval process that gives employees a pathway to legitimate access to AI tools. That pathway from unsanctioned to sanctioned is the operational promise that makes discovery worthwhile.

4.5 Manager's Checklist

Use this checklist to assess and advance your team's AI discovery posture:

25. Confirm whether your organization's IT team has activated AI domain categorization in DNS or web proxy logs and is reviewing those logs on a defined schedule.
26. Request a report from accounts payable or expense management on software subscription expenses from the past twelve months, and review it for AI tool purchases or reimbursements.
27. Ask legal or procurement to review your team's active SaaS vendor contracts for AI feature clauses, data training rights provisions, and recent terms-of-service updates.
28. Determine whether your organization has a formal AI tool registry, and if so, whether your team's tools are accurately represented in it.

29. Conduct a structured team discovery conversation with each direct report, using an amnesty-backed framing that separates disclosure from disciplinary review.
30. Document every AI tool your team uses, including tools that may be running inside approved platforms, and submit that documentation to the central registry or governance team.
31. Identify the tools your team considers most valuable, and begin the process of requesting formal governance review, either provisional approval or full sanction, before the next quarterly check-in.
32. Establish a standing agenda item in team meetings for AI tool updates, so that new tool adoption is reported through a predictable channel rather than discovered retroactively.

4.6 What Discovery Actually Requires

The discovery challenge in AI governance is not fundamentally a technology problem. It is a structural problem: the tools and processes that organizations built to track software assets were designed for a different era of tool adoption, and they do not naturally surface the web-based, employee-driven, vendor-embedded AI usage that characterizes the current moment. Closing the inventory gap requires layering technical detection methods with human-mediated disclosure programs, and it requires organizational credibility, the demonstrated commitment that disclosure is safe and that the information collected will be used to enable better work rather than to punish the workers who have already found better ways to do it.

A complete AI tool registry is not the end state of the discovery program. It is the enabling condition for everything else in an AI governance program. Risk assessment, policy development, vendor evaluation, compliance management, and cross-functional accountability all depend on knowing what is actually in use. Organizations that skip discovery and begin governance work with incomplete inventories will find themselves governing a subset of their actual AI exposure while the remainder continues to operate without visibility. The completeness of the inventory is the first governance variable a manager controls, and it sets the ceiling on everything that follows.

5 Governance That Gets Used

A global logistics company spent nine months developing what its legal team considered an exemplary AI governance policy. The document ran to thirty-two pages. It covered definitions, prohibited categories, approved vendor criteria, data classification requirements, incident reporting obligations, and a cross-reference matrix linking each policy provision to the relevant regulatory framework. It was thorough, well-sourced, and largely unread. Within four weeks of its release, a spot survey of business unit managers found that fewer than a third had read past the first five pages, fewer than one in ten could accurately describe the tool-approval process mandated by the policy, and the volume of unsanctioned AI tool deployments in the enterprise had not changed.

The policy had failed not because it was wrong about the risks, but because it was written in a way that made compliant behavior harder than non-compliant behavior. Finding the approved tool list required navigating three links from the policy document. Submitting a tool approval request required completing a fourteen-field form routed to a security review queue with an undisclosed processing time. Employees who needed an AI tool to meet a deadline that week did not wait for a process with no defined turnaround. They used what was available and hoped the policy would not be enforced. In most cases, it was not.

This chapter is a practical guide to building AI governance frameworks that achieve adoption by design policies specific enough to be actionable, flexible enough to accommodate legitimate use cases, and enforced through systems that are transparent and proportionate. The goal is to be a governance program that employees experience as helpful rather than obstructive, and that managers can apply without becoming policy experts. Getting there requires attention to policy architecture, authorization design, behavioral uptake mechanics, and the processes that keep governance current as the AI landscape evolves.

5.1 The Anatomy of Effective AI Policy

Effective AI policy is not primarily a legal document. It is a decision-support tool, a resource that helps employees and managers make good choices in real time without consulting a specialist. That framing has practical consequences for how policy is structured, what it covers, and how it is written. A policy that functions as a decision-support tool is short, specific, and written in the language of the people who must use it. A policy that functions as a legal liability shield is long, comprehensive, and written in the language of the counsel who drafted it. These two objectives are genuinely in tension, and the organizations with the most effective AI governance programs have made a deliberate choice to prioritize the former.

The decision to prioritize usability over comprehensiveness is not a sacrifice of legal protection. A policy that no one reads provides less legal protection than a policy that shapes actual behavior, even if the latter is shorter and less exhaustive. Courts and regulators evaluating corporate conduct in AI incidents will ask whether the organization had a governance framework and whether it operated in practice, not whether the framework was sufficiently lengthy. A ten-page policy that employees can apply in the moment is more defensible than a thirty-page policy that was filed and forgotten immediately.

Precision Over Comprehensiveness

The most common structural failure in AI governance policy is attempting to address every possible scenario in a single document. The result is a document that addresses nothing with sufficient clarity for practical application. Precision means identifying the specific decisions employees will face, determining whether this tool can be used for this task with this type of content, and providing guidance that resolves those decisions without requiring a compliance consultation.

Precision starts with use-case specificity. Generic prohibitions on the use of AI tools with sensitive data are not actionable guidance. Employees cannot assess whether their content is sensitive in the abstract; they can assess whether a patient name is in the file they are about to paste, or whether the client proposal they are about to summarize contains information marked confidential. The policy language that drives behavior is language tied to content types, workflow roles, and tool categories that employees recognize from their daily work. Abstract principles are training materials; specific standards are governance.

Organizations that have built precision-first policies typically begin with a use-case inventory rather than a list of principles. They ask: What are the ten most common ways employees in this organization want to use AI tools? What are the five uses that represent the highest risk? What does a useful answer look like for each of those questions? The resulting policy is organized around those use cases rather than around a theoretical risk taxonomy, and it is far more likely to be consulted and applied correctly at the moment of need.

Diagram 5.1: Policy Precision vs. Comprehensiveness

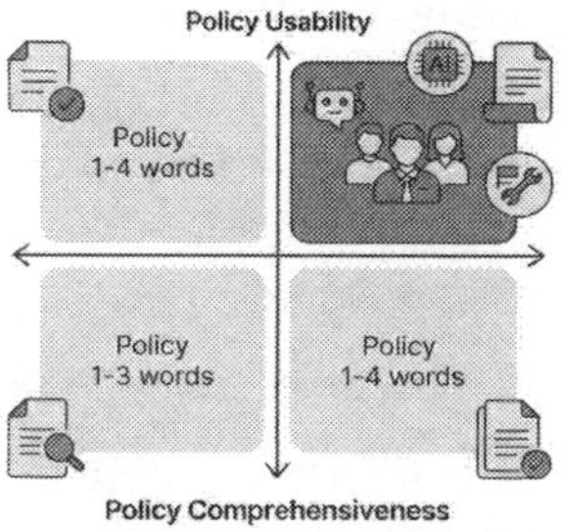

Balanced policy precision maximizes both employee comprehension and operational adherence.

Use-Case Classification

A use-case classification system is the organizing spine of a precision-based AI policy. Rather than asking employees to interpret abstract risk principles and apply them to specific situations, a classification system pre-maps common workflows to governance outcomes. The employee does not need to reason from first principles; they need to find their use case in the classification and follow the guidance it provides.

A functional use-case classification system covers the AI task types most common in the organization, drafting text, summarizing documents, generating code, analyzing data, answering questions, creating presentations, processing images, and maps each task type to the content sensitivity categories relevant to the organization. The intersection of task type and content category determines the governance outcome: permitted without restriction, permitted with specified controls, requires prior approval, or prohibited. A manager reviewing the classification system for their team should be able to answer, in under two minutes, what the governance position is for any combination of task and content that their team is likely to encounter.

Classification systems require maintenance, and that maintenance requirement is a feature rather than a bug. A system designed for maintenance forces the governance team to revisit classification decisions regularly, incorporating new use cases as they emerge and updating outcomes as the risk landscape changes. The alternative is that a comprehensive static policy document becomes obsolete without anyone noticing, because no one was reading it in the first place.

The Plain-Language Test

Before any AI policy provision is finalized, it should pass a plain-language test: can a business unit manager with no legal or security background read the provision and accurately predict what behavior it requires or permits, without asking for clarification? The test is deceptively demanding. Policy language that seems clear to the team that drafted it is often opaque to those who must apply it, because the drafters share a conceptual framework that the audience does not.

Administering the plain-language test requires actually testing policy language with representative members of the intended audience, not asking whether the language seems clear in the abstract, but presenting specific scenarios and asking managers to apply the policy and describe what it requires. Discrepancies between the intended and actual interpretations reveal where the language needs to change. Legal review of the revised language can follow; the sequencing matters. Writing for legal defensibility first, then testing for usability, produces a policy that passes legal review but fails in practice. Writing for usability first, then reviewing for legal sufficiency, produces a policy that works in both directions.

5.2 Acceptable Use Standards: A Tiered Framework

A tiered acceptable use framework is the operational core of an AI governance policy. It translates the classification system into concrete permissions and restrictions, organized by tier, and provides the reference structure that employees and managers consult when making tool and usage decisions. The tiered framework approach has become the governance design of choice for organizations with functioning AI governance programs because it scales with organizational complexity, is simple enough for employees to navigate, is specific enough to drive consistent decisions, and is structured enough to support enforcement.

The three-tier structure permitted, restricted, and prohibited is the most widely implemented because it matches the cognitive categories that most employees naturally apply when assessing whether a behavior is appropriate. More granular tier structures (four or five tiers) are technically possible but tend to produce confusion at the point of use because the distinctions between adjacent tiers become difficult to apply without expert interpretation. Three tiers, well-defined, provide sufficient specificity for most governance purposes.

Diagram 5.2: Three-Tier Acceptable Use Framework

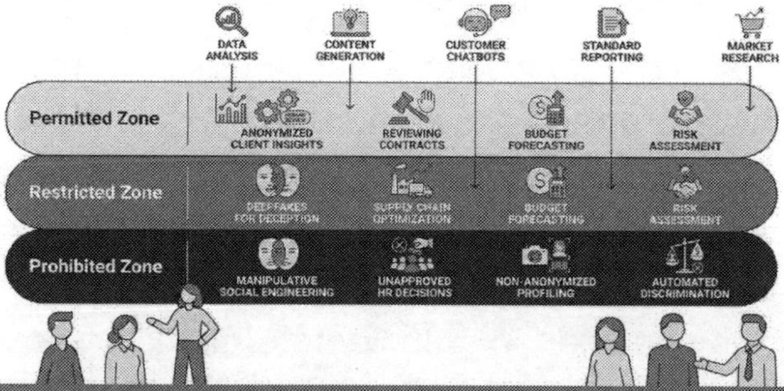

Caption: Consulting this framework guides employees toward the acceptable use of AI

Defining Permitted, Restricted, and Prohibited

Permitted use cases are those that the organization has evaluated and determined present acceptable risk under the existing governance controls, without requiring per-use authorization. Typical permitted categories include using approved AI tools to draft non-sensitive internal communications, summarize publicly available documents, generate boilerplate code for non-production environments, create initial outlines for presentations, and translate content that does not contain confidential information. The defining characteristic of a permitted use case is that the governance controls are already in place, the tool is sanctioned, the data involved is not sensitive, and no additional approval is needed for routine use.

Restricted use cases are those that present elevated risk but can be permitted under specified controls or with prior authorization. Restricted categories typically include using AI tools with internal but non-public business information, generating customer-facing content without human review, using AI tools to process data covered by contracts with non-disclosure provisions, and deploying AI outputs in automated workflows without human-in-the-loop checkpoints. The defining characteristic of a restricted use case is that the risk is manageable with additional controls specified in sufficient detail by the governance policy for the employee to implement them without seeking interpretation.

Prohibited use cases are those that the organization has determined present unacceptable risk regardless of controls, or that are prohibited by applicable law or regulation. Prohibited categories typically include processing regulated health information in non-healthcare-approved AI tools, submitting classified or export-controlled information to any AI tool without explicit security clearance, using AI tools to generate content that could be mistaken for official organizational communications without authorization, and using AI tools in ways that violate vendor contracts or data processing agreements. Prohibited categories should be short. A prohibited list that is so long as to be inoperable is an enforcement failure waiting to happen.

Role-Based AI Permissions

A critical limitation of a pure content-and-task classification system is that it treats all employees as equally positioned to make governance judgments about AI usage. They are not. A data scientist using AI for model development operates in a fundamentally different risk context than an HR coordinator using AI to draft internal job postings. The data scientist may legitimately need to work with complex, sensitive data in AI tools as a core professional function, subject to specialized controls. The HR coordinator needs clear, simple guidance covering a narrower set of use cases. A governance system that provides the same level of guidance to both is optimized for neither.

Role-based AI permissions address this by layering an employee role dimension onto the classification system. Rather than a single acceptable-use framework that applies to all employees, role-based permissions provide function-specific guidance calibrated to the actual AI use cases and risk levels relevant to each role category. Software engineers have a permissions profile that covers code generation, debugging assistance, and automated testing support. Finance analysts have a profile that covers data summarization and report drafting. Customer service representatives have a profile that explicitly addresses which tools may be used in customer interactions and which may not.

Role-based permissions are not significantly more complex to develop than a single universal framework. They are a structured elaboration of the use-case classification system, applied through a role filter that already exists in the organization's HR and access management infrastructure. The development effort is front-loaded, but the operational clarity it produces and the reduction in the number of individual employees who need to make governance judgment calls for which they are not qualified make that investment worthwhile for any organization of significant scale.

Sunset Clauses and Review Triggers

An AI governance policy that lacks a mechanism for its own revision will become dangerously outdated. The AI tool landscape changes at a pace that makes twelve-month policy review cycles, which are standard for most enterprise technology policies, inadequate for maintaining governance currency. A policy provision that is well-calibrated for the risk environment when it is written may be significantly over-restrictive or under-restrictive eighteen months later, depending on how the tool landscape and vendor terms have evolved.

Sunset clauses are policy provisions that automatically trigger review after a defined time period, requiring affirmative reauthorization rather than passive continuation. For AI governance policy, sunset periods of six months for the permitted use list and ninety days for the prohibited category list are appropriate. Those cadences ensure that governance decisions are revisited frequently enough to track meaningful changes in the AI landscape without creating an administrative burden that organizational leaders are unwilling to sustain.

Review triggers supplement sunset clauses by initiating policy review in response to specific events rather than elapsed time. Appropriate review triggers for AI governance include: the announcement of a major AI capability by a significant vendor; a regulatory guidance document or enforcement action relevant to the organization's sector; a security incident involving AI tools anywhere in the industry that reveals a risk category not addressed in the current policy; and a significant change in the organization's AI tool portfolio following a procurement decision. Trigger-based review allows governance to respond to events rather than waiting for a scheduled cycle that may be months away.

Diagram 5.3: Policy Lifecycle with Sunset and Triggers

A proactive policy management cycle including sunsetting and trigger-based reviews.

5.3 Decision Rights and Authorization Architecture

One of the most consequential design decisions in an AI governance framework is the authorization architecture: who has the authority to approve the use of AI tools, at what level of the organizational hierarchy, and under what circumstances. Organizations that centralize all AI authorization in a single review committee find that the committee becomes a bottleneck that employees route around. Organizations that decentralize authorization without clear standards find that local decisions are inconsistent and that the aggregate risk exposure is invisible at the organizational level. The goal is a decision-rights structure that distributes authority to the organizational level, where risk can be most accurately assessed, while maintaining centralized visibility into the decisions being made.

Authorization architecture is distinct from policy content. The policy describes what is permitted, restricted, and prohibited. The authorization architecture describes who can decide, at what organizational level, for which categories of decision. A well-designed authorization architecture makes compliance the path of least resistance, not because it is lenient, but because it puts the right decision authority in the right hands for each category of AI governance decision.

Who Can Approve a New AI Tool

Approval of a new AI tool is the most consequential authorization decision in the AI governance framework. It determines which tools enter the sanctioned environment and on what terms. The authorization structure for this decision must balance two competing risks: moving too slowly and creating the productivity gap that drives shadow AI adoption in the first place, or moving too quickly and onboarding tools without adequate evaluation.

A functional approval structure assigns different levels of approval authority based on the tool's risk tier. Low-risk tools that process only non-sensitive, non-regulated content with strong data-isolation terms can be approved at the business unit level by a manager or director with authority over the relevant function, subject to registration in the central AI tool registry and notification to IT and security. Medium-risk tools require review by a cross-functional team, including IT security, legal, and the business unit requester, with a defined turnaround time of no more than 10 business days. High-risk tools, those processing regulated data or requiring special security architecture, require executive-level approval with full security and legal review.

The turnaround standard deserves emphasis. The single most common driver of shadow AI adoption in organizations that have approval processes in place is an approval process with no defined turnaround time. Employees who submit a tool request and receive no response for weeks will adopt the tool anyway because the work demand does not pause for the approval queue. Publishing and holding to a turnaround standard, and tracking performance against it, converts the approval process from a symbolic gesture into a functional governance instrument.

Delegation Frameworks

Delegation frameworks allow AI governance authority to flow from a central governance body to functional leaders and line managers in a structured way, without requiring every governance decision to travel up and back down the organizational hierarchy. Effective delegation preserves organizational accountability for aggregate risk while enabling the operational speed that makes compliant behavior practicable.

A delegation framework for AI governance typically grants business unit leaders authority to approve AI tools in the low-risk tier within defined parameters, no regulated data, an approved vendor list, and no registration required, and establishes reporting obligations to ensure the central governance team maintains visibility into decisions made at the delegated level. The delegation is not unlimited; it is bounded by the parameters established by the central governance body, which define the conditions under which local approval is appropriate. Decisions that fall outside those parameters are escalated to the appropriate review level.

Delegation frameworks work best when the conditions for delegation are written as positive permissions rather than negative exceptions. A delegation framework that says business unit leaders may approve tools that meet all of the following criteria, followed by a specific list, creates clear guidance that managers can apply confidently. A framework that allows business unit leaders to approve tools, except in the following circumstances, creates an ambiguous residual category that managers will interpret inconsistently, producing exactly the type of governance inconsistency that the delegation framework is meant to prevent.

Diagram 5.4: Authorization Hierarchy for AI Tool Approval

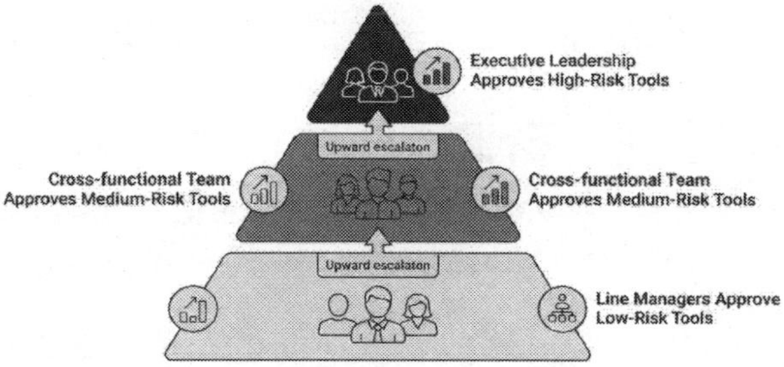

A structured process ensures responsible AI adoption by matching risk levels to appropriate decision-making authority.

Emergency Authorization Pathways

Emergency authorization pathways address the governance scenario that produces the greatest pressure toward non-compliance: the high-stakes, time-sensitive situation in which a business need is immediate, and the standard authorization process cannot respond quickly enough. A contract negotiation that needs AI-assisted analysis by tomorrow morning. A cybersecurity incident response that would benefit from an AI tool not currently in the sanctioned environment. A regulatory filing with a deadline that requires processing speed that approved tools cannot provide.

Emergency pathways must be designed carefully. A pathway that is too accessible becomes the standard route for impatient business units rather than a genuine emergency provision. A pathway that is too restrictive will be ignored, and emergency use will occur without any governance oversight. A functional emergency pathway requires a named approver, typically a CISO or CTO, who can reach a decision quickly, a documented request format that can be completed in under fifteen minutes, a defined authorization duration (typically twenty-four to seventy-two hours) rather than indefinite permission, and a mandatory post-authorization review that determines whether the tool should enter the standard approval process.

Emergency authorizations should be logged in the AI tool registry with their authorization basis and expiration date, and they should generate an automatic reminder to the requesting manager as the authorization approaches expiration. The log of emergency authorizations over time is itself a governance signal: a pattern of frequent emergency requests for the same tool categories indicates a structural gap in the organization's approved tool portfolio that the standard approval process should address proactively.

5.4 Policy Distribution and Behavioral Uptake

A governance policy that exists in a document management system and is unknown to the employees it governs is not a governance policy in any meaningful sense. It is a document. The distance between a policy document and a governance outcome is the distance between a published text and changed behavior. That distance is not closed by publishing the document in a more prominent location or by adding the policy to the annual compliance acknowledgment list. Behavioral uptake requires a distribution strategy that treats the policy as a communication challenge rather than a documentation task.

The organizations with the highest AI governance policy awareness and behavioral compliance share a common characteristic: they have invested in policy distribution and communication with the same seriousness they invested in policy content. They understand that the policy content is necessary but not sufficient, and that the governance program's operating effectiveness depends on the portion of the workforce that can accurately describe what the policy requires and apply it to their daily work decisions.

Getting Beyond the Checkbox

Annual compliance acknowledgment mechanisms, such as the digital signature that employees provide to confirm they have received and agree to abide by a policy, are a legal protection and a behavioral non-event. Research across compliance disciplines consistently finds that employees who have signed annual acknowledgments of the same policies each year retain almost none of the content. The acknowledgment creates a record; it does not create understanding or behavioral change. An AI governance policy that relies on annual acknowledgment as its primary distribution mechanism will achieve signature rates approaching 100% and behavioral uptake rates approaching 0%.

Getting beyond the checkbox means treating AI policy communication as an ongoing organizational behavior function rather than a periodic legal compliance exercise. It means making the policy accessible in the workflow context where governance decisions are made, embedded in the tool request process, linked from AI tool interfaces, and referenced in the manager's performance review conversation rather than in a separate compliance training module. It means providing scenario-based decision aids that translate policy language into questions employees can answer in the moment. And it means measuring whether the communication is working, not just whether it has occurred.

Behavioral uptake is also influenced by perceived fairness and clarity. Employees who believe the policy is sensible, that the restrictions it imposes are proportionate to the risks they address, and that the permitted uses reflect a genuine attempt to enable productive work, are far more likely to comply than employees who experience the policy as an arbitrary restriction imposed without explanation. The communication strategy should therefore address the rationale for the policy's key provisions, not just the provisions themselves. Employees who understand why a restriction exists are better positioned to apply it correctly in novel situations than employees who can only apply a rule they do not understand to the exact scenario covered in training.

Embedding AI Policy Into Onboarding

New employee onboarding is the highest-leverage point for AI governance policy distribution because it reaches every new hire at a time when behavioral norms are most malleable. An employee who enters the organization with a clear, accurate understanding of the AI governance framework will apply that understanding from day one. An employee who learns about AI governance through a year-end compliance module has already formed habits and possibly accumulated exposure over the months that have elapsed.

Embedding AI governance into onboarding requires more than adding a module to the existing compliance training stack. It requires positioning AI governance as part of the organizational culture conversation rather than the legal obligation conversation. The most effective onboarding integration presents AI governance alongside the organization's values on responsible technology use. It connects it to the specific tools and workflows the new employee will encounter in their role. An HR coordinator whose onboarding includes a concrete walkthrough of which AI tools are approved for HR functions, what restrictions apply to employee data, and how to request approval for additional tools will retain that information because it is immediately applicable to their work.

Managers play a critical role in onboarding AI governance norms. The manager who, in a new employee's first week, asks which AI tools they use and walks them through the organization's policy, is doing governance work that no centralized onboarding module can replicate. That conversation signals that AI governance is a real, applied expectation in this team, not a document sitting in a compliance folder. Manager involvement in onboarding AI governance is therefore a design requirement for the governance program, not an optional supplement.

Diagram 5.5: AI Policy Onboarding Integration

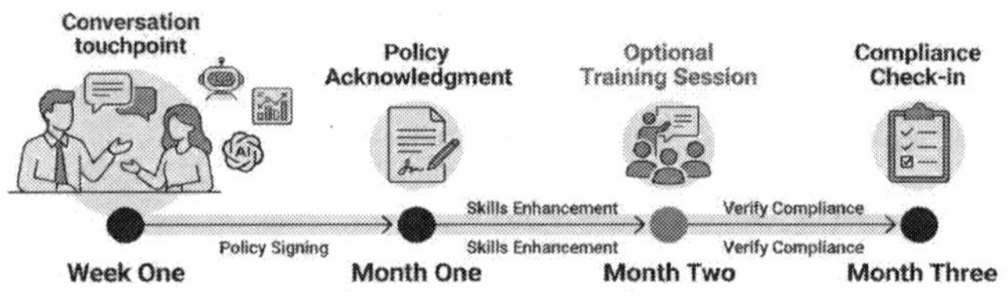

This structured process ensures effective AI policy adoption and ongoing governance integration.

Measuring Whether Policy Changes Behavior

Governance programs that do not measure behavioral outcomes will not know whether they are working until a significant incident reveals they were not. Measuring AI governance policy behavioral uptake is not technically difficult, but it requires agreement on what to measure and commitment to actually reviewing the data rather than filing it. The metrics that matter are not process metrics, such as how many employees completed training, how many signatures were collected, but outcome metrics that reflect whether the policy is shaping actual AI usage decisions.

Outcome metrics for AI governance behavioral uptake include: the ratio of AI tool requests submitted through the formal approval process versus new tools discovered through technical detection (a higher ratio indicates better policy awareness); the proportion of approved tool usage that follows specified controls as assessed through periodic spot checks; the frequency with which employees contact the governance team or legal department with AI usage questions (sustained high frequency indicates that the policy is not providing clear enough guidance at the point of use); and the prevalence of shadow AI usage as measured by the continuous discovery program over successive quarters.

These metrics should be reviewed on a defined schedule, at a minimum quarterly, by the governance body responsible for AI policy oversight. The review should explicitly ask whether the policy is achieving its behavioral objectives and trigger policy revision when evidence indicates it is not. A governance program that defines its own success criteria and reviews its own performance against them is operating as a real management function. A governance program that produces compliance artifacts and measures its own activity rather than its outcomes is performing governance theater. It will eventually lead to a significant incident to demonstrate the difference.

5.5 Keeping Governance Current

An AI governance policy that is not actively maintained will become less relevant faster than any other enterprise policy category. The AI tool landscape evolves continuously. Vendor terms of service change. New capability categories emerge that the original policy did not anticipate. Regulatory guidance updates create new compliance obligations. Employee workflows evolve as AI tools become more deeply integrated into business processes. A governance program without a formal mechanism to track and respond to those changes will find itself governing yesterday's AI environment. At the same time, the organization has moved to a significantly different one.

Keeping governance current is an operational discipline, not a crisis response. It requires assigning responsibility for surveillance and update initiation, establishing a cadence for scheduled reviews, and creating clear processes for both incremental updates (adding a new tool to the permitted list, adjusting a restriction based on updated vendor terms) and comprehensive revisions (rewriting the classification system following a major capability shift in the AI market). Both types of change will be necessary, and the governance infrastructure should support them without requiring a full mobilization of the governance committee for routine updates.

Trigger-Based Policy Updates

Trigger-based policy updates are the mechanism by which the governance framework responds to events rather than waiting for the scheduled review cycle. The trigger list should be maintained as a living document and distributed to the individuals responsible for monitoring in their respective domains: security for tool vulnerability disclosures; legal for regulatory guidance and enforcement actions; procurement for vendor terms changes and new tool acquisitions; IT for new AI capability launches by existing approved vendors.

When a trigger event occurs, the responsible monitor initiates a lightweight assessment: Does this event create a policy gap or render an existing provision inaccurate? If the answer is yes, the assessment proceeds to a recommended update, which is reviewed and approved through an abbreviated process that does not require convening the full governance committee. The abbreviated process should have a defined turnaround, typically five to ten business days, after which the update is either approved and implemented or escalated to the full governance body. This ensures that the governance framework stays current without requiring organizational mobilization for every minor update.

Stakeholder Input Loops

AI governance policy develops most effectively as a living dialogue between the governance function and the business units it serves. Stakeholder input loops are the formal mechanisms through which that dialogue occurs, structured opportunities for business unit leaders, operational managers, and front-line employees to provide feedback on governance policy, report practical difficulties in application, and surface emerging AI use cases that the policy has not yet addressed.

The most effective input loops are lightweight and embedded in existing organizational rhythms. A standing agenda item on the monthly IT leadership call asks whether governance policy has created any friction in the past thirty days. A brief survey of team managers at the end of each quarter, asking which AI tools their teams have requested and found unavailable through approved channels. A formal feedback channel from the tool approval process, collecting the reasons requestors provide when withdrawing requests, often reveals that the process itself was the obstacle rather than an informed decision that the tool was not needed.

Stakeholder input loops serve a purpose beyond policy improvement: they signal that the governance function is listening and that the program is designed to serve operational needs rather than to impose restrictions for their own sake. That signal matters for the organizational culture around AI governance. A workforce that believes governance is responsive to legitimate feedback is more likely to engage constructively by requesting tools rather than adopting them unilaterally, to report concerns rather than manage them privately, and to apply policy with judgment rather than seek technical workarounds.

Diagram 5.6: Stakeholder Input and Policy Feedback Loop

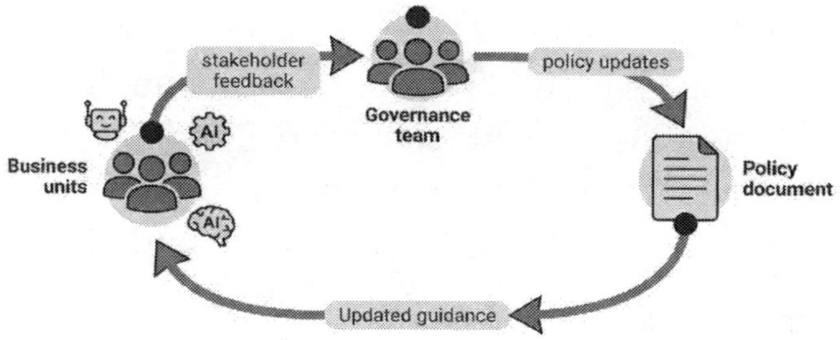

Iterative feedback loop ensures policy effectiveness and practical relevance.

5.6 Manager's Checklist

Use this checklist to assess whether your team's AI governance experience is functional or performative:

33. Locate your organization's current AI acceptable use policy and verify that you can find the three most important provisions for your team's AI workflows in under two minutes.
34. Test the policy's plain-language clarity by presenting two specific AI usage scenarios to a team member without a compliance background and asking them to apply the policy to determine whether the use is permitted.
35. Confirm that your team knows the tool approval process, specifically who submits the request, what information is required, and what the defined turnaround time is.

36. Review the permitted, restricted, and prohibited tier classifications for the use cases most common on your team, and identify any uses that fall into a gap the policy does not clearly address.
37. Assess whether any of your team's AI usage has changed since the policy was last updated, and if so, whether the policy remains accurate guidance for those uses.
38. Identify whether your team experienced any governance friction in the past quarter, or any case where the governance process slowed legitimate work, and report it through the appropriate stakeholder input channel.
39. Confirm that new employees on your team received AI governance orientation in their first week and that they can accurately describe what is permitted and how to request additional tools.
40. Review the emergency authorization pathway and confirm that you know how to use it and when it is appropriate.

5.7 Governance by Design

The standard against which an AI governance policy should be evaluated is not whether it is legally defensible in the abstract, but whether employees use it to make better decisions than they would make without it. That standard is demanding. It requires policy architects to think like the employee who is trying to finish a deliverable by Thursday, not like the legal team that needs to document organizational due diligence. It requires distribution strategies that reach people in the workflow context rather than in a training module context. And it requires a feedback infrastructure that treats employee difficulty with the policy as a design problem to be solved rather than a compliance failure to be enforced.

Governance that gets used is governance that earns its place in operational workflows by making the mission-aligned choice also the path of least resistance. When the tool approval process is faster than the workaround, employees request tools. When the permitted use guidance is clear enough to apply without consultation, employees apply it. When the feedback loop produces visible policy updates, employees engage with it rather than ignoring it. The pilot-to-production pathway for AI governance is not a technology deployment; it is a sustained investment in the organizational experience of compliance, treated with the same seriousness that good organizations bring to the employee experience of any other enterprise service.

6 Shared Accountability

Six months after a regional insurance company suffered a significant data incident involving an AI tool employees used to process policyholder claims data, the post-incident review found that, to almost no one's surprise, no single function owned the problem. IT had not detected the tool because it was accessed through personal devices. Legal had not reviewed the vendor terms because no one had submitted the tool for contract review. Compliance had not flagged the usage because it was not visible in any compliance monitoring channel. HR had not addressed the behavioral dimension because no one had classified it as an employee relations issue. And the business unit had reasonably, if incorrectly, assumed that using a productivity tool for internal work fell within the permitted scope.

This is the accountability failure pattern that characterizes the majority of significant AI governance incidents across industries. Shadow AI is not a technology problem, a legal problem, or a people problem in isolation; it is all three simultaneously, which is precisely why single-function ownership consistently fails. When IT owns AI governance, the program is technically rigorous but behaviorally thin: detection mechanisms work, but employee behavior does not change because the communication, training, and consequence functions that drive behavioral change lie elsewhere in the organization. When legal owns it, the policy is comprehensive but operationally disconnected: the governance framework is defensible in court. Still, it is not navigable by the employees who are supposed to apply it. When HR owns it, the culture conversation is rich, but the technical infrastructure and depth of risk assessment are absent.

This chapter examines the cross-functional accountability model required to manage unsanctioned AI effectively. It maps the specific ownership responsibilities of IT, legal, compliance, and HR, describes how those functions must interact through a governance body, and addresses the organizational dynamics that cause cross-functional accountability to break down, including the specific failure modes that lead to the outcome documented in the insurance incident above.

6.1 The Four-Function Map

Effective AI governance requires four institutional capabilities that no single organizational function fully possesses. Infrastructure and detection require technical expertise and systems access that live in IT. Regulatory interpretation and vendor contract evaluation require legal training and authority. Compliance monitoring and regulatory regime management require specialized knowledge that lives in compliance. And behavioral change, communication, and consequence management require the employee-relations tools and organizational authority that reside in HR. The four-function map is not an argument for committee governance of everything; it is a recognition that each function contributes a capability that the governance program cannot operate without, and that the program design must explicitly allocate those contributions rather than assuming they will emerge through informal coordination.

The four-function map also implies four distinct failure modes. When IT's contribution is absent, the governance program has no detection capability and no technical enforcement mechanism. When legal's contribution is absent, the program cannot accurately evaluate vendor terms, respond competently to regulatory inquiries, or assess the liability implications of AI incidents. When compliance's contribution is absent, the program cannot translate regulatory requirements into operational guidance for business units operating in specific sectors. And when HR's contribution is absent, the governance program has no mechanism for addressing the behavioral and cultural dimensions of AI adoption, which, as the insurance incident illustrates, are the dimensions most likely to determine whether the technical and legal frameworks are actually applied.

IT as Infrastructure and Detection

IT's role in AI governance spans four distinct responsibilities. The first is infrastructure deployment, building and maintaining the technical controls that enforce governance policy, including network monitoring for AI domain traffic, endpoint management configurations that restrict unauthorized AI tool installation where technically feasible, and API gateway controls that govern AI integrations with enterprise systems. These are configuration tasks that require IT expertise to implement correctly and operational ownership to maintain continuously.

The second responsibility is detection operations, which include running DNS log analysis, web proxy monitoring, and expense signal review, as described in the discovery chapter, and maintaining the AI tool registry as a technical database. Detection is not a one-time exercise; it is an ongoing operational function that generates the data inputs the governance program needs to maintain inventory accuracy and identify new compliance concerns as they emerge. IT owns the detection infrastructure and the operational cadence that keeps it running.

The third responsibility is to evaluate the security of new AI tools and vendors. When a business unit submits a tool approval request, IT security conducts the technical component of the review: assessing the vendor's security architecture, evaluating data handling and retention terms from a technical perspective, identifying integration risks with existing enterprise systems, and recommending security controls to accompany any provisional approval. This evaluation requires both security expertise and familiarity with AI-specific risk categories, such as model inversion attacks, prompt injection vulnerabilities, and training data extraction, that have emerged as the AI security threat landscape has matured.

The fourth responsibility is technical enforcement. When a tool is designated as prohibited under the governance policy, IT is typically responsible for implementing technical controls that make access to that tool more difficult from the corporate infrastructure, such as domain blocking, access controls, or endpoint restrictions. Technical enforcement is not a substitute for behavioral governance; determined employees will find technical workarounds, and technical blocking without communication can produce adversarial rather than cooperative employee responses. But technical enforcement signals organizational seriousness and raises the cost of non-compliant behavior in ways that pure policy communication cannot.

Diagram 6.1: IT's Four AI Governance Responsibilities

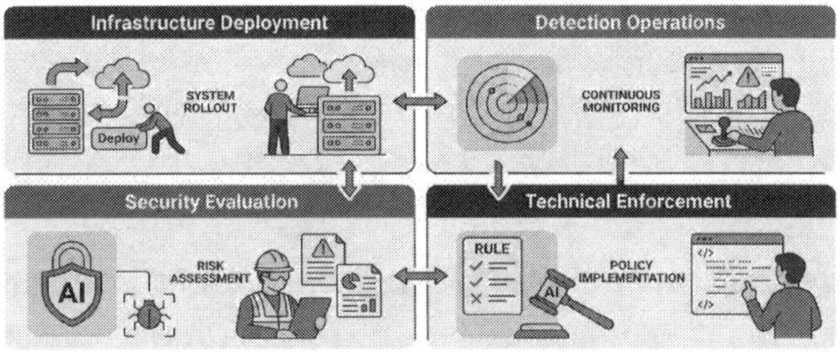

IT ensures AI systems are deployed securely, monitored continuously, and governed by established technical rules and policies.

Legal's Role in Vendor Evaluation and Incident Response

Legal's AI governance contribution centers on two primary functions: vendor assessment and incident response. Vendor assessment is the legal review component of the tool approval process, evaluating vendor contracts, data processing agreements, terms of service, and privacy policies to identify provisions that create unacceptable legal risk or conflict with the organization's regulatory obligations. This review is not interchangeable with IT's security evaluation; it requires legal training to identify the contractual and regulatory dimensions of risk that security assessments do not cover.

The vendor assessment scope for AI tools is broader than for most enterprise software because AI vendors frequently use terms-of-service provisions that are materially different from standard enterprise software agreements. Provisions governing whether user inputs can be used to train models, whether the vendor can share interaction data with third parties, how long interaction data is retained, and what the vendor's obligations are in the event of a breach are all consequential for the organization's legal exposure. They are frequently written in ways that are technically accurate but easy to misread by non-lawyers. Legal's assessment provides the authoritative interpretation that other functions need to make informed approval decisions.

In incident response, legal's role shifts to managing the immediate and downstream legal consequences of an AI-related security or compliance failure. This includes assessing notification obligations under applicable breach-notification laws, managing responses to regulatory inquiries, evaluating litigation exposure, and advising on public communication. AI incidents frequently involve multiple regulatory regimes simultaneously. A breach affecting patient data processed through an AI tool may trigger HIPAA breach notification, state breach notification laws, and FTC oversight all at once. Legal must coordinate across those regimes at operational speed, leaving no time for the deliberate pace of normal contract or policy review.

Compliance as the Regulatory Translation Layer

Compliance occupies a distinctive position in the AI governance function map because its primary contribution is to translate rather than execute. The compliance function does not implement technical controls, negotiate contracts, or manage employee relations. Its value lies in its ability to interpret the regulatory environment applicable to the organization's specific sector and translate that environment into operational guidance that IT, legal, HR, and business units can apply without becoming regulatory experts themselves.

For AI governance, regulatory translation means converting the requirements of applicable frameworks, such as HIPAA for healthcare, FedRAMP and FISMA for federal contractors, GDPR for organizations with EU data subjects, state-level AI regulations for organizations in covered jurisdictions, and sector-specific guidance from relevant regulatory bodies, into specific operational requirements that the governance program must address. These requirements affect vendor selection (which vendors are authorized to process regulated data), access controls (which employees may use which AI tools with which data categories), audit trail requirements (what documentation must be maintained for AI-assisted decisions), and incident response obligations (what must be reported to whom and within what timeframe).

The compliance function is also responsible for monitoring the regulatory horizon, tracking emerging AI-specific legislation, regulatory guidance documents, and enforcement actions that may require updates to the governance program. The AI regulatory environment is evolving faster than most compliance teams have encountered; guidance from the FTC, the EEOC, and sector-specific regulators on AI use has been frequent and consequential. Organizations that lack a compliance function with a dedicated AI regulatory monitoring responsibility will consistently be surprised by requirements they should have anticipated, converting governance program updates from planned improvements into crisis responses.

Diagram 6.2: Compliance as Regulatory Translation Layer

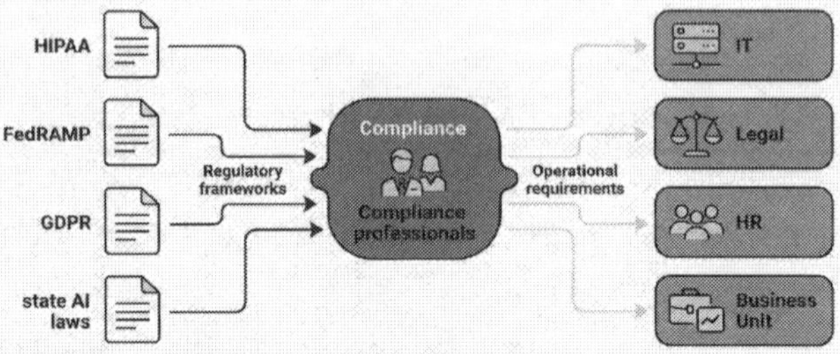

The compliance function translates complex **regulatory frameworks** into actionable operational requirements for various business units.

6.2 HR's Underutilized Role in AI Governance

Human resources is the least-utilized function in most AI governance programs, and this underutilization represents one of the most consequential design gaps in the field. AI governance programs that lack HR integration have diagnostic and detection capabilities but no behavior-change infrastructure. They can identify shadow AI use, assess its legal and technical risks, and draft a policy prohibiting it. Still, they cannot consistently change the behavior that produces it, because behavior change at an organizational scale is an HR competency, not a technology or legal one.

The behavioral dimension of shadow AI governance is not peripheral to the governance challenge; it is the governance challenge. The risk does not originate in the AI tools themselves; it originates in the decisions that employees make every day about which tools to use, what content to process through them, and whether to seek authorization before adopting a new workflow. Organizational culture, management norms, performance incentives, and the perceived consequences of compliance versus non-compliance shape those decisions. All of those factors are HR's domain.

Employee Relations and Behavioral Dimensions

The employee relations dimension of AI governance encompasses the interpersonal and organizational dynamics that determine whether governance policies are taken seriously and applied consistently. In practice, this means addressing three behavioral realities that are present in virtually every organization deploying AI tools: the prevalence of manager complicity in shadow AI adoption, the role of peer norms in normalizing unauthorized tool use, and the relationship between job insecurity and employees' willingness to use AI tools in ways that may not be fully sanctioned.

Manager complicity is the most consequential behavioral dynamic in shadow AI governance. When team managers know that their staff is using unauthorized AI tools because the output quality of deliverables has improved, because they can see the tools in use during team calls, or because employees have disclosed their usage informally, and choose not to address it, they effectively authorize the behavior. Their inaction signals that the governance policy is not a real organizational expectation. Reversing that signal requires explicitly engaging managers in the governance program, providing them with clear guidance on their accountability, on what they are expected to do when they observe unauthorized tool use, and on the support the organization provides to help them have those conversations constructively.

Peer norm dynamics are harder to address directly, but matter significantly for long-term governance effectiveness. When AI tool use is widespread within a team and perceived as the norm, individual employees who attempt to comply with governance restrictions experience social friction; they work more slowly or produce less polished output than colleagues using unauthorized tools. The implicit message is that compliance is a competitive disadvantage. Breaking that dynamic requires governance programs that provide sanctioned alternatives quickly enough that compliance does not feel like a productivity sacrifice, and that communicate governance enforcement broadly enough that non-compliance feels like a genuine risk rather than a norm everyone secretly accepts.

Performance Management and AI Disclosure

The intersection of performance management and AI governance is an area where HR's governance contribution is both critical and delicate. If employees believe that disclosing AI tool use will negatively affect their performance evaluations because it reveals they have been using unauthorized tools, or raises questions about whether AI-assisted work outputs are truly their own, they will not disclose. Disclosure programs built on amnesty commitments alone are insufficient if the performance management framework does not reinforce those commitments. Managers who evaluate employee work without acknowledging or understanding the role of AI tools in producing it create an implicit pressure toward both non-disclosure and unsanctioned usage.

HR's role is to design performance management frameworks and evaluation criteria that are coherent with the AI governance program. This means explicit guidance to managers on how to evaluate AI-assisted work, how to discuss AI tool usage in performance conversations, and how the organization's stance on AI use as a professional capability should be reflected in competency frameworks. An organization that has published an AI governance policy but whose performance review forms and manager evaluation criteria do not refer to AI usage has created a structural inconsistency that employees will resolve in favor of the system, thereby affecting their compensation and career progression.

Performance management integration also means revisiting what outcomes the organization rewards at the individual contributor level. If speed and volume of output are the primary performance metrics, employees will use whatever tools most efficiently produce them, and they will not stop to verify whether those tools are authorized. If mission-aligned delivery and operational clarity are rewarded, employees have a different optimization target that is more compatible with governance compliance. HR controls the reward structures that shape those optimization targets, and those structures are as consequential for AI governance as any technical control.

Diagram 6.3: HR's Three AI Governance Contributions

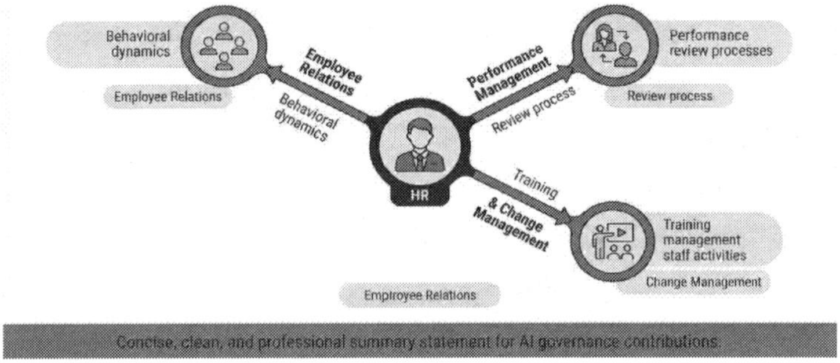

HR as the Change Arm of Governance

Change management is the fourth and most strategically important dimension of HR's AI governance contribution. Governance programs that are technically sound and legally complete frequently fail because they are deployed into organizational culture without the change management infrastructure needed to translate the program's requirements into new behavioral norms. Employees do not change how they work because a policy document was updated. They change how they work because their managers are modeling different behavior and because training has equipped them with the competencies to work differently. After all, the tools they need to work in the new way are available, and because the organizational culture signals that the new way is the expected standard.

All of those change levers belong to HR. Training program design and delivery, manager competency development, organizational communication, and culture assessment are HR functions. An AI governance program that deploys without HR change-management support is being deployed into an organizational environment that has not been prepared to implement it. The gap between the policy and the workforce's actual behavior will be filled by whatever cultural defaults were already in place, which, in most organizations, are permissive toward technology adoption and skeptical of compliance overhead.

HR's change management contribution to AI governance should begin before policy is deployed, not after. The culture assessment that identifies where current employee attitudes toward AI tool use and governance compliance are concentrated, the manager readiness evaluation that determines where coaching investment is most needed, and the communication design that positions the governance program as an enabler rather than a restriction are all pre-deployment activities. Organizations that engage HR in AI governance design rather than governance rollout achieve better behavioral outcomes because the program architecture incorporates the behavioral environment from the beginning, rather than discovering it as an obstacle after launch.

6.3 Designing the Cross-Functional Governance Body

A four-function accountability map requires a governance body that can coordinate across those functions, make decisions when functions disagree, and maintain a consistent institutional position on AI governance that does not vary by which function currently has the most organizational momentum. The governance body is not a replacement for functional accountability; it is the coordination mechanism that ensures the four functions' contributions are integrated into a coherent program rather than into four parallel, potentially contradictory governance efforts.

The design of the governance body itself is a governance decision. Getting it right requires attention to membership, decision-making authority, meeting cadence, and mechanisms to prevent the governance body from becoming an obstacle rather than a resource. Organizations that have attempted AI governance through loose inter-functional coordination and informal relationships between individuals in IT, legal, compliance, and HR, without a formal governance structure, consistently report that coordination breaks down under pressure, particularly during incidents when speed is critical and each function's natural institutional priorities conflict.

AI Steering Committee Architecture

The cross-functional AI governance body, commonly called an AI Steering Committee or AI Governance Council, should be structured to make operational decisions at the speed required by the AI governance environment. That means clear membership, defined decision-making authority, and a meeting cadence that creates a consistent governance rhythm without consuming organizational capacity disproportionate to the governance work.

Membership should include a senior representative from each of the four core functions: a CISO or senior IT security leader, a deputy general counsel or senior legal counsel with privacy and technology expertise, a chief compliance officer or senior compliance director, and a chief people officer or senior HR business partner as well as representation from the business units most heavily engaged in AI adoption. The business unit representation is critical and frequently omitted. A governance body composed exclusively of control functions will produce governance that is technically sound but operationally disconnected; the business-unit perspective grounds governance decisions in the operational reality experienced by employees and managers.

Decision authority must be explicit. The governance body should have clear authority to approve tools in the medium-risk tier, to update permitted and restricted use classifications without requiring board-level approval, and to issue interim governance guidance in response to emerging risks without waiting for a full policy revision cycle. Decisions reserved for executive leadership, high-risk tool approvals, responses to regulatory enforcement actions, and significant policy changes affecting enterprise-wide practices should be defined clearly, so that the governance body understands what it can decide and what it must escalate. A governance body without defined authority will either overstep and create organizational resistance or understep, allowing governance gaps to persist.

Diagram 6.4: AI Steering Committee Architecture

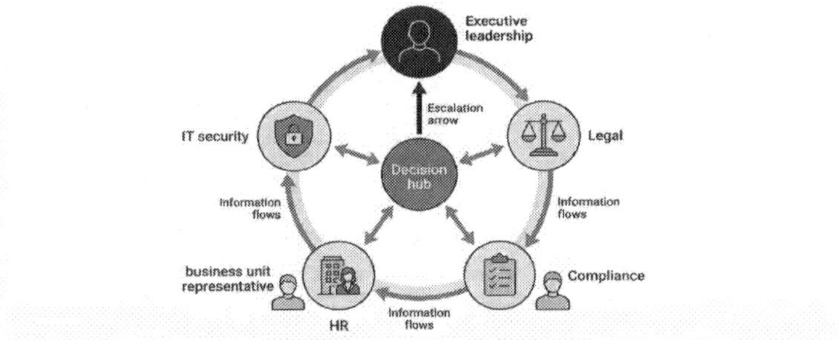

A circular steering committee architecture to foster collaborative AI governance and decision-making.

Decision Escalation Paths

Escalation paths formalize how decisions move from the operational governance level to the executive level when the stakes or complexity exceed what the steering committee is authorized or equipped to resolve. Clear escalation paths prevent two common failure modes: governance body paralysis, where the committee defers difficult decisions indefinitely rather than escalating them appropriately; and escalation flooding, where every AI governance question gets referred to executive leadership regardless of actual stakes, overwhelming executive bandwidth, and undermining the committee's operational credibility.

The escalation criteria should be written specifically enough to guide application without requiring case-by-case judgment in every instance. Tool approvals involving regulated data always escalate. Decisions with estimated financial exposure above a defined threshold always escalate. Situations involving a regulatory inquiry or litigation hold always escalate. Decisions where two or more committee members formally object to a proposed course of action are escalated by right, allowing dissenting members to invoke the escalation path without requiring majority approval. Those criteria, written down and agreed to by the governance body and executive sponsors before they are needed, prevent the ad hoc escalation negotiation that occurs under pressure when the governance framework did not anticipate a specific scenario.

Escalation paths should also include return communication requirements. When a matter escalates to executive leadership and a decision is made, the decision and its rationale should be communicated back to the governance body with enough specificity that the committee can apply the same reasoning to similar future decisions. Escalation that disappears into executive deliberation and returns only as a directive, without explanation, does not build the governance body's competency; it trains the committee to defer upward rather than to develop the judgment to decide.

Avoiding Committee Calcification

Cross-functional governance bodies tend to calcify over time. The initial energy of establishing the committee gives way to routine; membership becomes fixed rather than refreshed; meeting agendas focus on status reporting rather than consequential decisions; and the governing body gradually transforms from a decision-making institution into a coordination meeting that an email chain could replace. Calcification is the natural trajectory of governance bodies that do not actively manage against it.

The most effective prevention mechanism is a periodic governance effectiveness review, a structured assessment, conducted annually, of whether the governance body is making the decisions it was designed to make, whether those decisions are being made at the right organizational level, and whether the membership and process structure are still fit for the governance challenge the organization faces. The governance environment for AI tools is not static, and a committee architecture designed for one stage of the organization's AI adoption may be poorly suited to a later stage, when AI is more deeply integrated, and governance questions are more complex.

Membership rotation is the second line of prevention. Governance body members who have held their seats for multiple years develop working assumptions about the organization's AI risk profile that may not be updated as the landscape evolves. Bringing in new members, either as permanent additions or as rotating representatives from business units, introduces perspectives that prevent the committee from governing based on the environment that existed when it was established rather than the one that exists now.

6.4 Accountability Breakdowns and Prevention

Cross-functional accountability structures fail in predictable ways. Understanding those failure modes before they occur is the most practical form of prevention available, because the dynamics that produce accountability breakdowns tend to operate gradually and are easy to rationalize individually, even as they systematically undermine the governance program. The four failure modes described below are documented across multiple industry sectors and are not unique to AI governance; they appear wherever significant organizational risks span multiple institutional functions without clearly unified ownership.

Diagram 6.5: Four Accountability Failure Modes

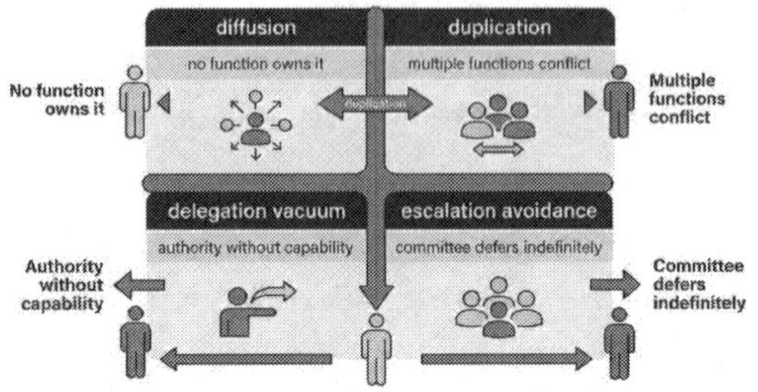

This chart Four accountability failure modes defers backs to the business dysroblems.

The "Somebody Else's Problem" Failure Mode

The most common accountability breakdown in cross-functional governance is the diffusion of responsibility, the collective assumption, implicitly shared by all functions, that another function is taking primary responsibility for a governance dimension that is, in fact, unowned. This failure mode is not the result of bad intentions; it is a structural consequence of accountability systems that define functional responsibilities without explicitly mapping the spaces between functions. When every function believes its own responsibilities are covered and assumes that adjacent responsibilities belong to someone else, the spaces between functions become governance voids.

In AI governance, the most common diffusion points are: the ongoing monitoring of vendor terms changes (legal assumes IT is tracking; IT assumes legal is reviewing); the behavioral follow-up on discovered shadow AI usage (compliance assumes

HR is handling; HR assumes the business unit is addressing it); and the update of the AI tool registry following governance body decisions (IT assumes the governance coordinator is maintaining it; the governance coordinator, if one exists, assumes IT is updating it in their system of record). Each of these diffusion points represents a real governance gap that will eventually produce a consequential failure.

Prevention requires explicit assignment of ownership for every governance activity that crosses function boundaries. The assignment must designate a particular individual, for example, "the deputy general counsel," instead of referencing a function such as "legal." It should also specify a distinct delivery method, such as a mandatory registry update within five business days following the committee's decision. Accountability assigned to a function is accountability that everyone in that function can assume someone else is handling. Accountability that is assigned to a named individual cannot be diffused without a visible act of reassignment.

Incident Response Across Function Lines

AI governance incidents, data exposures, regulatory inquiries, employee conduct situations involving AI tools, or public disclosures of shadow AI usage are the moments when cross-functional accountability is tested most severely, because they require rapid, coordinated action across all four functions simultaneously. IT must assess the incident's technical scope and implement containment measures. Legal must evaluate notification obligations and manage external communications. Compliance must assess regulatory reporting requirements. HR must address employee conduct issues and manage internal communications with the workforce.

These four parallel workstreams must be coordinated in real time, under pressure, by individuals who may not have previously worked together at the required intensity. The quality of that coordination is almost entirely determined by whether the organization has a pre-defined AI incident response protocol that assigns responsibilities, establishes communication channels, and defines decision authorities before the incident occurs. Organizations that attempt to improvise cross-functional incident response at the moment of crisis consistently produce fragmented, inconsistent, and legally consequential responses.

The AI incident response protocol should be developed as a joint product of all four functions, reviewed and tested through tabletop exercises before a real incident occurs, and maintained as a living document, updated following each real or simulated incident. The tabletop exercise is particularly important because it reveals the decision ambiguities and communication gaps that the written protocol does not anticipate. A protocol that has never been exercised is a document; a protocol that has been exercised is a governance instrument.

Post-Incident Review as Governance Improvement

The post-incident review is the mechanism by which AI governance incidents become governance improvements rather than isolated events. Without systematic post-incident review, each incident is managed, resolved, and forgotten, leaving the governance program with the same vulnerabilities that produced it. In a systematic review, each incident is analyzed to document what failed, why it failed, and what governance changes would reduce the probability of recurrence.

A functional post-incident review covers four questions: What governance gap allowed the incident to occur? Which function's accountability was implicated, and was it unclear, unexercised, or simply unavailable given the speed of events? What changes to governance programs would address the gap? And what is the timeline and ownership for those changes, with a verification mechanism? Post-incident reviews should occur within thirty days of resolution, and the implementation status of recommendations should remain a standing agenda item until confirmed complete. Reviews that produce recommendations never implemented are governance theater that erodes the program's credibility.

6.5 Accountability Under Regulatory Pressure

The accountability dynamics of cross-functional AI governance shift when an organization faces active regulatory scrutiny. The normal operating rhythm, governance body meetings on a defined cadence, function-specific responsibilities managed in parallel, and escalation paths reserved for significant decisions are replaced by a compressed, high-stakes coordination requirement that simultaneously exposes every gap in the accountability structure. Understanding how regulatory pressure reshapes governance accountability and designing the governance structure to remain coherent under that pressure is the final design requirement for cross-functional AI governance.

Regulatory inquiries involving AI usage arrive through multiple channels, depending on the sector and the nature of the inquiry. Federal agencies conducting oversight of AI use in government contracting may issue formal information requests with tight response deadlines. Healthcare regulators investigating privacy incidents may conduct on-site reviews requiring real-time document production and employee interviews. State attorneys general investigating AI-enabled consumer harms may issue civil investigative demands that require coordinated responses from legal, compliance, and business unit personnel. Each of these inquiry types activates the cross-functional governance structure in different ways and at different levels of urgency.

Diagram 6.6: Cross-Functional Response Under Regulatory Pressure

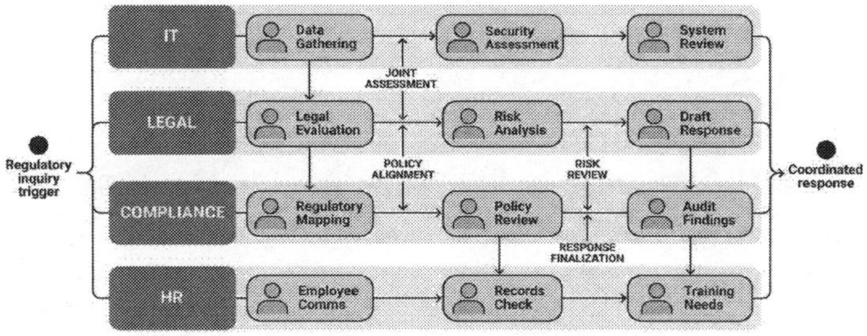

Cross functional response, esonse in action (team raropensed tonms) underpnancing the regulatory pressure.

Federal Oversight Environments

Federal agencies and government contractors operating under federal oversight frameworks face AI governance accountability requirements that are more prescriptive and more directly enforceable than those in most private sector contexts. The FedRAMP authorization framework, FISMA compliance requirements, and agency-specific AI governance policies issued by the Office of Management and Budget and individual agency heads create a layered accountability environment in which AI tool approval, usage monitoring, and incident reporting obligations are formally specified rather than left to organizational discretion.

In federal oversight environments, the cross-functional governance body has a formal compliance function in addition to its operational governance function. It must be able to produce documentation of its governance processes, tool approval decisions, and incident response activities in response to oversight requests. That documentation requirement has architectural implications for the governance program. All governance body decisions must be formally recorded, tool registry entries must be maintained with sufficient detail to support oversight reviews, and incident response protocols must produce the documentation artifacts oversight agencies will expect.

Federal contractors who have allowed shadow AI to proliferate, allowing employees to use unapproved AI tools with government data, face a particularly acute accountability exposure when federal oversight arrives. The AI tool registry will be the first document requested, and the absence of a complete, accurate registry is itself an accountability finding that can trigger deeper inquiry. Federal oversight is, in this sense, a forcing function for the completeness of governance programs that private-sector organizations encounter less directly, which makes the federal experience a useful governance design reference even for organizations not directly subject to federal oversight.

Healthcare Governance and AI Incident Custody

Healthcare organizations face AI governance accountability requirements shaped primarily by HIPAA and state health privacy laws covering AI-processed health information. The incident custody question, which function is responsible for an AI governance incident involving protected health information, and what are the obligations from discovery through required notifications, is the most practically consequential accountability question in healthcare AI governance.

HIPAA breach notification requirements are triggered by unauthorized access to, acquisition of, use, or disclosure of protected health information. An employee using an unsanctioned AI tool to process clinical notes, patient communication drafts, or claims data containing PHI creates a potential HIPAA breach the moment that data leaves the covered entity's control environment. The breach assessment, the four-factor test evaluating the nature of the PHI, the unauthorized person, whether PHI was acquired or viewed, and the extent of mitigation, is a legal and compliance function. But the underlying facts required to conduct that assessment are owned by IT and HR.

Healthcare AI governance must include a clear incident-custody protocol that assigns responsibility for breach-assessment coordination to a specific named role, typically the chief privacy officer or privacy counsel, while establishing the information-gathering obligations that IT and HR must fulfill. The custody protocol should be exercised through tabletop simulation before it is needed, and it should explicitly address the time pressure imposed by HIPAA's required notification timeline. Healthcare organizations with clear, exercised incident-custody protocols respond to AI-related PHI incidents significantly faster than those that improvise in the moment of crisis.

6.6 Manager's Checklist

Use this checklist to assess the cross-functional accountability health of your organization's AI governance program:

41. Confirm that your organization has identified named accountable individuals, not just functions, for each core AI governance responsibility: detection, legal review, regulatory monitoring, behavioral change management, and registry maintenance.
42. Verify that a cross-functional AI governance body exists, meets on a defined schedule, and has explicit authority to make tool approval decisions in the medium-risk tier without requiring executive-level escalation.
43. Identify the escalation criteria in your governance body's operating charter and confirm that they are specific enough to guide application without case-by-case negotiation.
44. Assess whether HR is meaningfully integrated into the governance program, specifically, whether HR has contributed to manager guidance on AI-related accountability conversations and whether performance management frameworks reference AI governance expectations.

45. Confirm that a documented AI incident response protocol exists, names specific function responsibilities and decision authorities, and has been exercised through at least one tabletop simulation in the past twelve months.
46. Review the most recent governance body meeting record and confirm that it reflects consequential decisions rather than status updates that the body is making governance choices, not coordinating information sharing.
47. Ask each functional representative what their primary AI governance accountability is and verify that their answers are consistent with the formal governance design rather than individually interpreted.
48. Confirm that post-incident review outputs are being tracked to completion and that the governance body has a mechanism for verifying that recommended changes have been implemented.

6.7 The Accountability Architecture That Holds

Cross-functional AI governance works when accountability is specific, visible, and designed to hold under organizational pressure. Specific means named individuals with defined responsibilities and delivery mechanisms, not functions with broadly described roles. Visible means governance body decisions documented in records that named individuals can access, registry entries that reflect the actual governance status of every tool in operation, and incident response documentation that creates a contemporaneous account of what each function knew and did. Designed to withstand pressure, with escalation paths that function during incidents, not just during routine operations, and post-incident review processes that convert operational failures into institutional learning.

The shared accountability model creates a governance architecture with a clear coordination mechanism, clear function-specific accountabilities, and clear escalation paths for decisions that transcend any single function's authority. That architecture, maintained with the same operational discipline applied to any other critical business process, produces the governance coherence organizations need to manage the workflow-level impact of AI adoption without surrendering visibility or control. Operational clarity about who owns what is not bureaucratic overhead; it is the structural prerequisite for a governance program that performs when performance is required.

7 From Rogue to Recognized

The procurement team at a regional insurance company had been using an AI-powered document summarization tool for eleven months when the CISO's office noticed it during a routine vendor contract review. The tool had been adopted informally by three analysts who found it through a product review newsletter. By the time the security team had completed DNS log analysis, the tool had processed over 4,000 documents, including actuarial models, policyholder correspondence, and draft litigation summaries. No one had done a vendor assessment. No data processing agreement existed. The tool's privacy policy permitted the provider to use uploaded documents for model improvement, subject to opt-out settings that no one on the team was aware of. The CISO was left with a choice: prohibit the tool immediately and absorb the productivity disruption or conduct an emergency evaluation and decide whether the value justified the conditions under which it could be retained.

That choice and the structured process required to make it well are what this chapter addresses. Most shadow AI governance frameworks are built around detection and prohibition. They answer the question of what to stop. They do not answer the harder question of how to evaluate whether a prohibited tool should become permitted, or what steps convert a rogue deployment into a recognized, monitored, and controlled capability. Organizations that lack that pathway make every discovery a crisis and every decision a binary: ban it or ignore it. Neither option reflects sound governance.

7.1 Why Recognition Matters

The instinct to prohibit is understandable. When a tool has been operating without authorization, the safest-seeming response is to shut it down, close the exposure, and reissue the policy, reminding employees that unauthorized software is forbidden. That response has the virtue of decisiveness, but it has the defect of incompleteness. It does not address why the tool was adopted in the first place, it does not capture the workflow need that drove the behavior, and it does not offer the employee a sanctioned path forward. Within weeks, a functionally equivalent tool or the same tool under a different account re-emerges.

The case for a structured sanctioning pathway rests on three realities. First, many shadow AI tools provide genuine productivity value that the enterprise has not matched through approved channels. If the tool is eliminated without replacement, the workflow reverts to slower, lower-quality manual methods, or the employee finds another unsanctioned substitute. Second, the cost of sanctioning a low-risk tool is often lower than the cost of sustained enforcement against a tool people are determined to use. Third, organizations that develop a reliable fast-track evaluation process build a credibility dividend: employees learn that bringing AI tools forward through governance channels leads to useful outcomes rather than automatic refusal, which encourages future disclosure. Prohibition without an alternative pathway creates a culture of concealment.

This chapter presents the evaluation architecture that allows governance teams to move from discovery to a sanctioned, monitored, or formally prohibited disposition in a structured way. The goal is operational clarity: every discovered tool should have a documented path to resolution, and every resolution should produce a record that informs future decisions.

DIAGRAM 7.1: SHADOW AI SANCTIONING DECISION TREE

7.2 The Evaluation Threshold

Risk-Benefit Scoring for Unsanctioned Tools

Every evaluation begins with a risk-benefit scoring exercise that produces two numbers: a risk score reflecting the tool's potential to cause harm, and a value score reflecting the productivity or quality benefit the tool delivers. Neither number is meaningful in isolation. A high-risk, high-value tool requires more rigorous controls but may still be worth sanctioning, provided appropriate safeguards are in place. A low-risk, low-value tool may not be worth the governance investment and can be quietly replaced with a nudge toward an approved alternative.

Risk scoring for shadow AI tools should address five dimensions. Data sensitivity examines what categories of information the tool processes: does it handle PII, financial records, health data, or attorney-client communications? Vendor trust evaluates the provider's privacy posture, terms of service, data processing agreements, security certifications, and organizational history. The scope of exposure considers how many users have adopted the tool and how much data has already passed through it. Regulatory intersection maps whether the tool's use cases touch compliance-sensitive domains such as HIPAA-covered workflows, FedRAMP-bounded systems, or GDPR-scoped data subjects. Integration depth assesses whether the tool operates in isolation or has access to broader enterprise systems through APIs, credentials, or browser-level access to authenticated sessions.

Value scoring requires input from the users who have adopted the tool. The evaluation team should conduct brief structured interviews asking three questions: What work does this tool perform that you could not do as well with approved tools? How much time does the tool save per week? What would happen to your output quality or volume if the tool were removed today? These answers give the governance team a grounded view of the workflow-level impact the tool is delivering, which is essential for negotiating sanction terms or specifying what the approved alternative must provide.

Diagram 7.2: Risk-Benefit Scoring Matrix

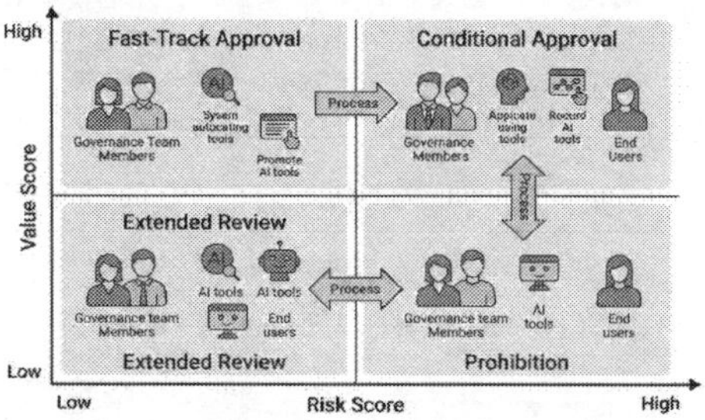

This matrix assists the governance team in prioritizing and approving AI initiatives based on their risk and benefit profile.

Vendor Assessment for Consumer-Origin AI

Shadow AI tools are disproportionately consumer-origin products, designed and marketed for individual users, that have been adopted at scale within enterprise environments. Consumer-grade AI products typically have terms of service that were written for individual accountability, not organizational liability. They may include training data provisions that permit user-uploaded content to improve the model, retention policies that keep data long after the session ends, and privacy settings that are opt-out rather than opt-in. Enterprise procurement teams evaluating consumer-origin AI must apply a different lens than they would to enterprise software designed with organizational compliance in mind.

A vendor assessment for consumer-origin AI should cover six areas. Terms of service review focuses on data ownership, training data clauses, and whether enterprise-tier accounts with different data handling terms are available. The security posture review examines the SOC 2 Type II certification status, penetration testing history, vulnerability disclosure policies, and incident response track record. Data processing agreement availability determines whether the vendor will sign a DPA that satisfies the organization's regulatory obligations and internal data governance standards. Sub-processor transparency reviews which third parties the vendor uses for infrastructure, model hosting, or analytics, and whether those sub-processors are subject to equivalent contractual obligations. Jurisdictional analysis identifies where data is processed and stored, which becomes critical for organizations subject to cross-border data flow restrictions. Model training policy clarifies whether the vendor's enterprise product uses customer data to train or fine-tune its models, and what opt-out or data isolation mechanisms are available.

Not every vendor will pass all six checks. The governance team's role is not to apply a binary pass-fail test but to document gaps, determine whether contractual negotiation can close them, and assess whether the residual risk is acceptable given the tool's value. That documented judgment is what distinguishes a mature evaluation process from an ad hoc decision.

Security and Privacy Review in Compressed Timelines

One of the practical obstacles to sanctioning shadow AI tools is a mismatch in timelines. Standard enterprise software procurement cycles run for 60 to 90 days. Shadow AI tools that are already in active use create urgency. Every day the tool remains in an evaluated-but-unsanctioned state is a day when the organization cannot enforce consistent data-handling standards. The governance framework must include a compressed review track for tools that present either high usage volume or bounded risk profiles.

A compressed security and privacy review should target a ten-business-day cycle from discovery to preliminary disposition. Day one through three focus on data classification: determine what data types the tool has processed and what the exposure surface looks like. Days four through six focus on vendor assessment using the framework described above, drawing on existing certifications and publicly available documentation rather than waiting for full vendor questionnaires. Days seven and eight focus on risk scoring and value scoring, incorporating direct input from the user group. Days nine and ten focus on preliminary disposition: either provisional approval with conditions, referral for extended review, or formal prohibition with rationale. Every outcome is documented in the AI tool registry, regardless of disposition.

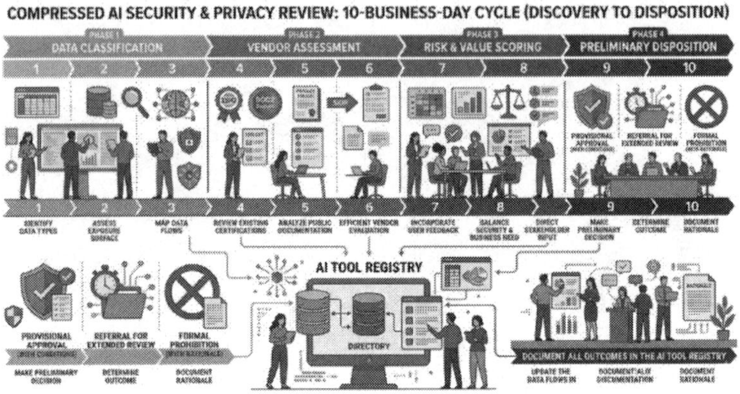

7.3 Fast-Track Authorization

The 72-Hour Provisional Approval Framework

For tools that present a clearly bounded risk profile, typically those that process only non-sensitive public or internal-use data, have no access to enterprise systems, and come from vendors with established enterprise-tier offerings, a 72-hour provisional approval framework provides the governance team with a mechanism to enable continued use while the full evaluation proceeds. Provisional approval is not full sanctioning. It is a time-limited, condition-bounded authorization that allows the organization to continue receiving business value from the tool while the review is completed.

Provisional approval requires four conditions to be met. First, the tool must fall below the data sensitivity threshold established in the risk-benefit scoring framework: no PII, no protected health information, no confidential financial data, and no attorney-client privileged material. Second, the users must agree in writing to restrict their use of the tool to the designated low-sensitivity use cases during the provisional period. Third, the tool's primary account holder must be identified and designated as the governance contact for the evaluation period. Fourth, a full evaluation timeline must be established and communicated, with a hard deadline by which the tool either receives full sanctioning or must be discontinued.

The 72-hour timeline is achievable when the governance team uses pre-built evaluation templates rather than starting each review from scratch. A well-designed template covers vendor assessment checklist items, data sensitivity classification criteria, user interview questions, and the risk-benefit scoring rubric. Teams that invest in building these templates before the first shadow AI discovery save weeks of process-building time during the first real evaluation.

Diagram 7.3: Provisional Approval Workflow

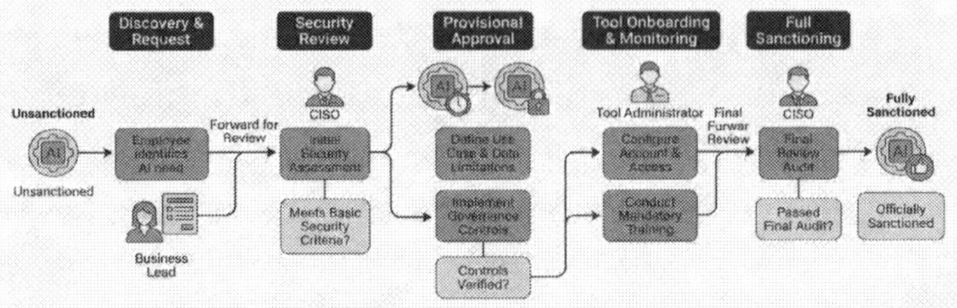

Caption: Miniematic identified approval workflow in denerriea/oeanpiatios, evencIven officially sanctioned.

Lightweight Risk Acceptance for Low-Sensitivity Use Cases

Not every shadow AI tool requires a full evaluation architecture. A subset of tools that operate entirely on non-confidential data, perform tasks such as grammar checking, format conversion, or general web research summarization, and do not connect to enterprise systems can be processed through a lightweight risk-acceptance pathway. This pathway bypasses the full vendor assessment and the compressed review cycle in favor of a structured risk-acceptance document signed by the relevant business-unit leader.

The lightweight risk acceptance document records six items: the tool name and vendor, the specific use cases for which the tool is accepted, the data categories that are explicitly out of scope, the name of the responsible business unit leader, the acceptance date, and the next review date. This document does not constitute full sanctioning; it is a formal acknowledgment that the business unit has evaluated the tool's use cases, determined them to be low-sensitivity, and accepted residual risk within the bounds of their authority. The risk acceptance pathway ensures that governance overhead is proportionate to actual risk, which is one of the design principles that make governance frameworks usable in practice.

Conditional Approval: Time-Boxed Authorization

Many tools fall between the clear pass and clear fail of the evaluation process. They offer genuine value, they have vendor terms that are not ideal but are negotiable, and they present bounded but real risk that requires ongoing monitoring. For this category, conditional approval, a time-boxed authorization attached to specific conditions, is the right disposition.

Conditional approval specifies five elements. The authorization scope defines which teams and use cases are covered, and the approval covers data types. The conditions enumerate the requirements that must remain satisfied for the authorization to remain valid: these may include a signed DPA with the vendor, active user training on data-handling requirements, and quarterly review by the governance team. The monitoring obligations specify the logging, alerting, and periodic auditing that the IT team must perform during the authorization period. The review date establishes when the conditional approval expires and must be formally renewed or converted to full sanctioning. The escalation trigger defines the conditions under which the authorization is immediately suspended: a material change in vendor terms, a security incident involving the tool, or evidence of use outside the defined authorization scope.

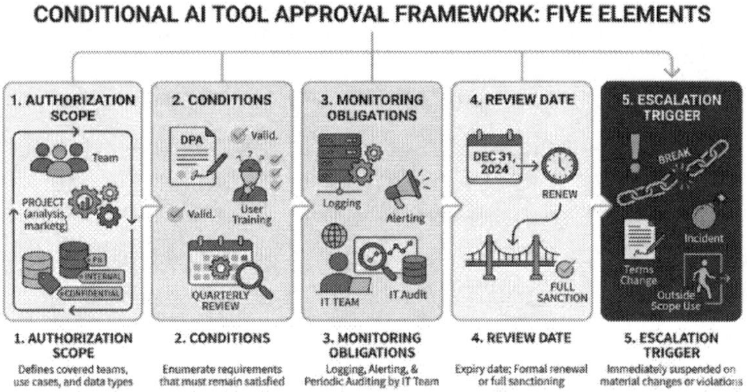

Conditional approval is not a shortcut in governance. It is a recognized risk management instrument that allows organizations to capture value from tools that are not ready for full integration while maintaining defined accountability for their use. Every conditional approval must be reflected in the AI tool registry with its conditions and review date clearly documented.

7.4 The Integration Transition

Data Handling Requirements and Onboarding Checklist

When a shadow AI tool receives full sanctioning, the organization must complete an integration onboarding process to close the gap between informal use and governed deployment. This onboarding does not require the tool to be technically integrated into enterprise systems; many sanctioned tools remain standalone web applications or browser-based services, but it does require a structured transition that establishes consistent data handling practices, user awareness, and monitoring obligations.

The data handling requirements that apply at the time of sanctioning should be documented in the tool's entry in the AI tool registry and communicated to all current and future users. These requirements specify what data categories are permitted for use with the tool, what categories are explicitly prohibited, how outputs from the tool must be labeled or attributed, whether outputs require human review before use in official documents or client deliverables, and what retention or deletion obligations apply to outputs generated by the tool. These requirements should mirror the organization's existing data classification and handling standards and be extended to apply to AI-generated content.

The onboarding checklist outlines the procedural steps required before the tool is formally added to the approved list. Checklist items include: execution of a data processing agreement with the vendor, configuration of enterprise-tier account settings including any available data isolation or opt-out features, completion of user training on the tool's data handling requirements, designation of a tool administrator who is responsible for monitoring and governance communications, creation of the monitoring plan, and logging of the completed onboarding in the AI tool registry with all supporting documentation attached.

User Communication at Sanctioning

The way an organization communicates a sanctioning decision shapes whether the governance process earns credibility or squanders it. Employees who used the tool during its unsanctioned period are watching: they want to know whether governance produced a fair outcome, whether their workflow need was taken seriously, and whether the result makes their work better or worse. Communication that treats sanctioning as a bureaucratic closure, a form letter announcing the tool is now approved, misses the opportunity to demonstrate that governance operates in the organization's interests.

Effective sanctioning communication has four components. First, it acknowledges the legitimate need the tool addresses and validates the employees who discovered its value. This is not about congratulating rule-breaking; it is about recognizing that unsanctioned adoption often reflects a real gap in the approved tool portfolio. Second, it explains the conditions attached to approval in plain language so users understand what they can and cannot do with the tool. Third, it names the tool administrator and provides a clear channel for reporting issues or asking questions. Fourth, it establishes the timing of the next review and invites ongoing feedback. This communication should come from the business unit leader, not the CISO's office, to reinforce that governance is a shared responsibility and not a surveillance function.

Diagram 7.4: Integration Onboarding Sequence

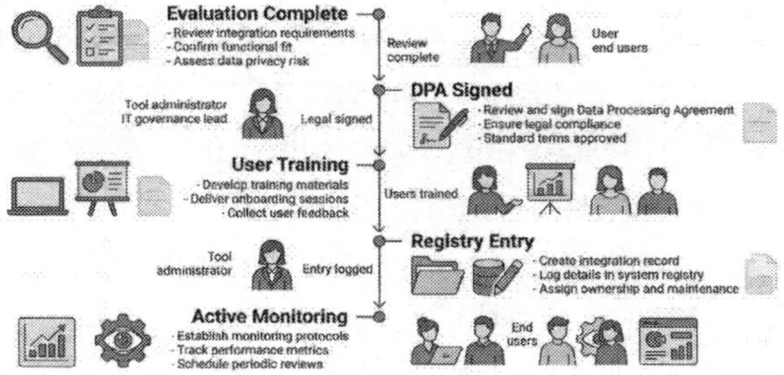

This flowchart illustrates the essential steps for successfully integrating new tools within the organization.

Monitoring Obligations After Approval

Sanctioning a tool does not end the governance responsibility; it begins a new phase of it. Post-approval monitoring closes the loop between authorization and ongoing oversight, ensuring that the tool continues to be used under the conditions established at the time of sanctioning and that any material changes to vendor terms or tool capabilities are detected and reviewed.

Post-approval monitoring covers four areas. Usage monitoring tracks which users are accessing the tool and at what volume, using available vendor reporting or network-level logging to detect significant changes in scope. Data type monitoring looks for signals that the tool is being used with data categories outside its authorized scope. This may involve periodic user surveys, anomaly detection in network traffic, or DLP policy alerts triggered by file types associated with sensitive data categories. Vendor monitoring tracks changes in the vendor's terms of service, data processing practices, ownership structure, or security posture, with a designated team member responsible for reviewing vendor communications and annual certification renewals. Conditional review manages the review dates established at the time of sanctioning, ensuring that each tool's conditional approval is formally re-evaluated rather than allowed to persist indefinitely due to administrative inertia.

7.5 Managing Tools That Cannot Be Approved

Enforcement Architecture for Prohibited Categories

Some shadow AI tools will not pass the evaluation. They may use training data clauses that the organization cannot accept, process data in jurisdictions that violate regulatory requirements, lack the security certifications required by the organization's third-party risk standards, or present capabilities that are inappropriate for the enterprise context, regardless of other considerations. For these tools, the governance framework must provide a clear, proportionate, and consistently applied enforcement architecture.

Enforcement architecture for prohibited AI tools has three layers. Prevention uses technical controls to block access to known prohibited endpoints, including DNS filtering for consumer AI platforms that are categorically prohibited, DLP policies that detect and block uploads of sensitive data categories to unapproved AI services, and browser extension management policies that prohibit the installation of unapproved AI browser tools on managed devices. Detection maintains the ability to identify prohibited tool use that circumvents prevention controls, including reviewing web proxy logs for traffic patterns associated with known prohibited services, monitoring expense reports for subscriptions to prohibited tools, and maintaining an active program of periodic manager-led team disclosure. Response establishes a consistent process for addressing discovered prohibited tool use: a documented conversation with the user, a review of what data was exposed, a determination of whether incident response obligations are triggered, and a decision on whether remedial or disciplinary action is warranted under HR policy.

Enforcement architecture must be proportionate. The goal is not to treat every policy violation as a terminable offense; it is to maintain consistent standards, close exposures, and build the kind of organizational culture in which employees understand that AI governance is a real operational discipline, not a paper exercise. Heavy-handed enforcement of low-risk violations damages the trust that governance programs depend on to function. Light enforcement of high-risk violations signals that the policy is not serious.

Communicating Denial Without Underground Adoption

When a tool evaluation concludes with a prohibition decision, communicating that decision is as important as the decision itself. A poorly communicated denial drives adoption underground. The employee who received the prohibition notice has an unmet workflow need that led to the original adoption. If the denial does not address that need, the employee will find another way to meet it.

Prohibition communication should follow a four-part structure. First, it should state the decision clearly and without bureaucratic hedging: the tool has been evaluated and cannot be approved. Second, it should explain the specific reasons for denial in terms the employee can understand: not legal jargon about data processing agreements, but plain language about what the tool does with data that the organization cannot accept. Third, it should provide an approved alternative that addresses the same workflow need, with enough specificity that the employee can begin using it immediately. If no approved alternative exists at the time of denial, the denial communication should include a commitment to evaluate alternatives within a defined timeline. Fourth, it should invite the employee to provide feedback on whether the approved alternative meets their need, creating a feedback loop that informs future tool evaluation priorities.

Diagram 7.5: Prohibition and Alternative Pathways

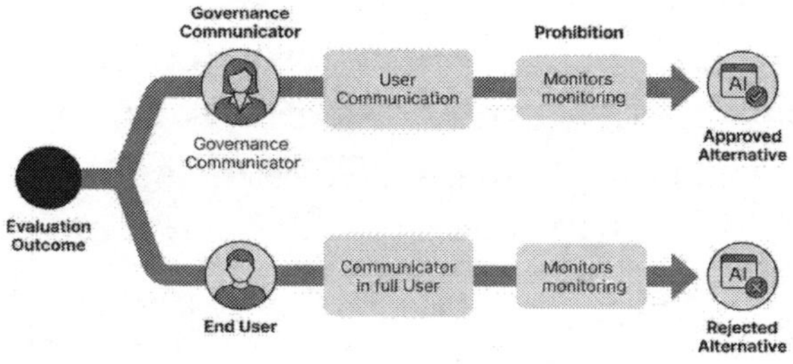

The diagram illustrates the governance steps for identifying alternatives during AI prohibition decisions.

Providing Approved Alternatives

The most effective component of a prohibition decision is the approved alternative. Organizations that invest in building and maintaining a curated catalog of approved AI tools, organized by use case rather than product category, can respond to prohibition decisions with immediate alternative recommendations. This catalog should be a living document, updated as new tools are evaluated and sanctioned, and accessible to all employees as a self-service resource.

The approved AI tool catalog should be organized around the tasks employees are trying to accomplish. Categories might include document summarization, draft generation, code assistance, data analysis, translation, and meeting transcription. For each category, the catalog should list the approved tools, the data types each tool is authorized to process, any restrictions on use, and the point of contact for access provisioning. When a prohibition decision references the catalog, the employee has a concrete resource to consult rather than an abstract instruction to use approved alternatives.

Building the catalog requires proactive tool evaluation, not waiting for shadow AI discovery to drive the evaluation agenda. Governance teams should maintain a forward evaluation queue that anticipates the tool categories employees are most likely to seek, based on productivity trend data, industry reporting, and input from the manager network. A governance program that can approve tools faster than employees feel compelled to adopt unsanctioned ones has achieved one of its most important operational goals.

Diagram 7.6: Approved AI Tool Catalog Architecture

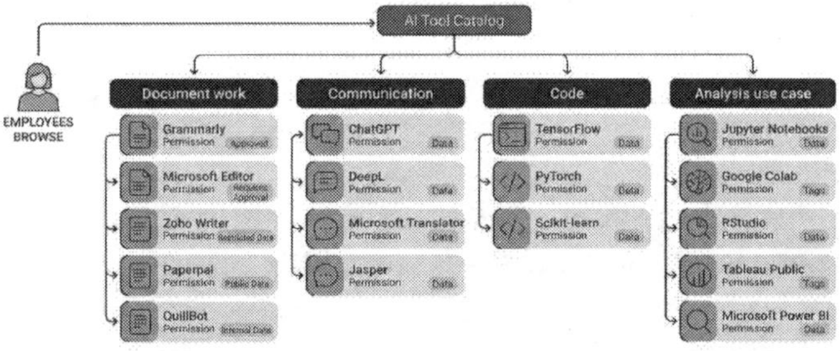

A structured repository of vetted and approved AI tools categorized for employee access and efficient business use cases.

7.6 Manager's Checklist

- Ensure your team knows how to report a discovered shadow AI tool. The reporting pathway should be visible and frictionless.
- When a shadow AI tool is discovered on your team, do not unilaterally prohibit or endorse it;

route it through the governance evaluation process.

- Participate in value scoring interviews for tools your team has adopted: the evaluation team needs your perspective on workflow impact.
- Review provisional approval conditions with your team and confirm they understand which use cases are authorized during the evaluation period.
- Confirm that your team's AI tool administrator has been designated and knows their responsibilities.
- After a sanctioning decision, deliver the user communication personally rather than forwarding a generic governance notice.
- When a tool is prohibited, provide the approved alternative recommendation in the same conversation; never let a denial stand without an alternative.
- Track your team's conditional approval expiration dates and initiate the renewal review at least three weeks before the deadline.
- Conduct quarterly informal check-ins with your team on AI tool usage. These conversations surface new shadow adoption before it becomes a formal discovery.
- Provide feedback to the governance team on whether approved alternatives are meeting your team's workflow needs.

7.7 What This Chapter Established

The pathway from rogue to recognized is not a reward for policy violation; it is a recognition that prohibition without evaluation and prohibition without alternatives fails as a governance strategy. Organizations that build structured evaluation processes, fast-track authorization frameworks, and approved alternative catalogs create a governance environment in which employees have a credible reason to bring AI tools forward rather than conceal them. That shift from a culture of concealment to a culture of disclosure is the most important outcome a shadow AI governance program can produce.

The evaluation architecture presented in this chapter is deliberately tiered. Lightweight risk acceptance for low-sensitivity tools, provisional approval for tools that need time-bounded authorization during review, conditional approval for tools that require ongoing monitoring, and full sanctioning for tools that pass the complete evaluation cycle. Each tier is calibrated to match governance overhead to actual risk, a design principle that makes governance programs usable rather than merely thorough.

When governance teams approach every shadow AI discovery with a structured process, a clear disposition pathway, and a commitment to providing approved alternatives, they transform what might otherwise be a crisis response into an operational capability. That capability, the ability to evaluate, authorize, monitor, and, when necessary, prohibit AI tools consistently and transparently, is the operational foundation of responsible AI governance. The next chapter addresses the cultural and communication work that converts that operational capability into an enterprise-wide behavioral norm.

8 The Enablement Pivot

Three months after a large professional services firm issued its first formal AI acceptable use policy, adoption of the firm's approved AI tools had not increased; it had decreased. The CISO's communications had been thorough, the policy language had been reviewed by legal, and the training modules had been deployed through the LMS with completion tracking. Every procedural box had been checked. What had not been considered was how the policy felt to the people subject to it. The predominant tone of every communication about AI governance had been cautionary. The opening lines of the policy document warned employees about data exposure, regulatory liability, and the consequences of unauthorized use. The training modules spent thirty percent of their runtime on enforcement and discipline. The message employees received, even though no one had intended to send it, was that AI was dangerous, the organization was not ready for it, and the safest choice was to avoid it entirely.

The result was a governance program that reduced visible AI use but did not reduce actual AI use. Shadow adoption continued. Approved tool usage remained flat. And the employees who were using AI most productively, those with the domain expertise to apply it well and the judgment to use it responsibly, were the ones most likely to be discouraged by a governance framework that felt adversarial. The firm had built a compliance program when what it needed was an enablement program.

This chapter addresses the pivot that governance programs must make to sustain themselves: from a posture of restriction and caution to one of disciplined enablement. That pivot is not about lowering standards or ignoring risk. It is about redesigning how the organization talks about AI, how it builds employee capability, how it distributes judgments across the workforce, and how it creates a communication environment in which responsible AI use is recognized as a professional competence rather than a compliance obligation.

8.1 Why Enablement Is a Governance Strategy

The most durable governance programs in enterprise history, whether for information security, financial controls, or data privacy, share a characteristic that is easy to overlook in their early stages: they succeeded because employees eventually internalized their logic, not merely their requirements. When employees understand why a control exists and agree that it serves a legitimate purpose, they exercise judgment in ambiguous situations in ways that align with governance intent. When employees view governance as an external constraint imposed by a function they distrust, they comply when observed and find workarounds when not observed. That dynamic is exactly what plays out with shadow AI.

Restriction-only governance creates a particular failure mode that is expensive to diagnose because it appears compliant on the surface. Employees complete required training, acknowledge policy updates, and route approved use cases through authorized tools. Meanwhile, the high-stakes, high-value AI applications, the ones where governance matters most, continue to operate in the shadows, precisely because the governance framework has not given employees the confidence or authority to bring them into the light. Enablement governance breaks that dynamic by giving employees something the restriction model cannot offer: a reason to want to work within the system.

The enablement pivot begins with a straightforward reframe: AI governance exists to make AI useful to the organization in a way that is sustainable, legally defensible, and consistent with the organization's obligations to its clients, employees, and regulators. That reframe has to be genuine, not a marketing campaign layered over a prohibition-heavy policy. It requires governance teams to ask whether the tools they have approved are actually useful, whether the approval process is fast enough to compete with unsanctioned alternatives, and whether the people responsible for governance communicate in a way that signals helpfulness rather than enforcement. When the answer to those questions is yes, the enablement narrative writes itself.

Diagram 8.1: Restriction vs. Enablement Governance Postures

This diagram contrasts the impact of restrictive and enabling governance on AI adoption and compliant usage.

8.2 Reframing the Internal AI Narrative

Leadership Messaging That Feels Like Enablement

The organizational narrative about AI is set, consciously or not, by the most senior leaders who communicate about it. When the first communication employees receive about AI governance comes from the general counsel, the chief compliance officer, or the head of information security, the frame is already set: AI is a risk to be managed. That frame is not wrong; it is incomplete. It tells employees what the organization fears about AI without telling them what the organization believes AI can do for the people who work there.

Leadership messaging that enables rather than merely restricts begins with a statement of organizational intent: what the enterprise is trying to accomplish with AI, why AI capability matters to its competitive position and service quality, and what role employees play in building that capability responsibly. This statement should come from the CEO or a senior business leader, not exclusively from the functions responsible for governance. When business leadership is the primary voice on AI strategy, and governance functions are positioned as the enabling infrastructure rather than the restrictive authority, the organizational narrative shifts. AI becomes something the organization does intentionally, and governance becomes the framework that enables that use.

The practical test for leadership messaging is whether an employee reading the communication feels more capable of using AI well or more anxious about using it at all. Communications that lead with risk inventories, enforcement consequences, or regulatory exposure fail the test. Communications that lead with what AI can accomplish for the employee's specific work, followed by the framework that makes that use responsible, pass it. Governance teams that want to influence leadership messaging should draft the communications themselves and offer them for review, rather than waiting for communications they will then have to correct reactively.

The Chief AI Officer Communication Role

Organizations that have appointed a Chief AI Officer or an equivalent role responsible for enterprise

AI strategy have a significant structural advantage in the enablement pivot. The CAO is positioned to advance the AI narrative in a way neither the CISO nor the compliance function can: as a proponent of responsible AI use, rather than as a regulator of AI risk. The CAO's communication role is to be the organizational voice that makes AI governance feel like an investment in capability rather than a constraint.

The CAO communication cadence should include four recurring elements. Regular updates on the organization's AI tool portfolio, what has been sanctioned, what is under evaluation, and what is being actively explored, give employees a current picture of the organization's AI capability. Spotlights on internal AI use cases that delivered measurable value demonstrate that compliant use produces real results, not just compliance records. Transparent discussion of tools that were evaluated and not sanctioned, with clear explanation of why, builds credibility with employees who want to understand the governance logic rather than accept its conclusions. Invitations for employee input on AI tool needs signal that the governance agenda is driven in part by the people it serves, which is both organizationally accurate and strategically important.

In organizations that do not yet have a CAO, the communication role should be shared between the senior business leader most responsible for AI strategy and the CIO or CISO, with deliberate attention to ensuring that the business leader carries the enabling frame. At the same time, the technology and risk functions provide the governance infrastructure that makes it credible.

Storytelling With Internal Case Studies

The most persuasive evidence that responsible AI use produces real value is a story about someone in the organization using AI responsibly and achieving a result that matters. Internal case studies brief, factual accounts of how employees or teams used sanctioned AI tools to improve their work are among the most effective communication tools available to governance programs. They are credible because they come from peers, specific because they describe real workflows, and motivating because they show what compliant AI use actually looks like in practice.

Building an internal case study library requires a process for identifying, capturing, and communicating these stories. Managers are the best source of case study nominations: they see which AI-assisted workflows are producing visible results in their teams. The AI Champion network, described in detail later in this chapter, is another source. Case studies should be brief, three to five paragraphs, and follow a consistent structure: the challenge the employee or team faced, how they used an approved AI tool, the result they achieved, and what they would tell a colleague about using AI responsibly. The governance team should review them to confirm they describe tool use that complies with policy, and then share them through internal communication channels, all-hands meetings, and the intranet's AI resource center.

Diagram 8.2: Internal Case Study Distribution Flow

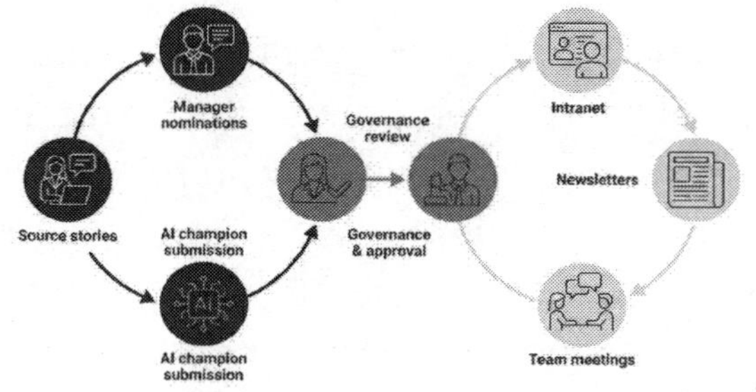

This diagram illustrates the standardized workflow for nominating, reviewing, and distributing internally generated AI case studies across the organization.

8.3 AI Literacy as a Governance Instrument

What AI Literacy Means for Non-Technical Leaders

AI literacy is frequently described as though it were a single capability; either you understand AI, or you do not. In practice, AI literacy is a layered concept with varying requirements across organizational levels. For a data scientist, literacy means understanding model architecture, training dynamics, and evaluation metrics. For a non-technical business leader, literacy means something more practical and more immediately applicable: the ability to recognize AI-generated content, understand the conditions under which AI outputs are reliable, formulate effective prompts for approved tools, evaluate AI-assisted recommendations with appropriate skepticism, and recognize when a decision requires human judgment that AI cannot substitute.

Non-technical leaders are the most important AI literacy constituency in the organization, and they are the constituency most commonly underserved by existing AI training programs. Programs designed by technical teams tend to start with how AI works rather than what it means for the work the manager is actually responsible for. The result is training that feels irrelevant to the business context, that managers deprioritize under workload pressure, and that leaves them less able to exercise the AI judgment their teams need. Governance programs that want to build genuine AI capability in the management layer must design literacy programs from the manager's perspective, starting with the decisions and workflows where AI is already relevant to their specific function.

For non-technical leaders, the practical dimensions of AI literacy include three core capabilities. Prompt quality judgment is the ability to recognize when a prompt is likely to produce useful output versus when it is too vague, too complex, or too sensitive to yield reliable results. Output evaluation is the ability to review AI-generated content with the same critical lens applied to any other analytical input, checking for factual plausibility, internal consistency, alignment with known context, and absence of hallucinated specifics. Risk recognition is the ability to identify situations in which AI-generated outputs carry governance implications: when an output is used to inform a decision with regulatory consequences, when sensitive data categories are introduced into a prompt, or when a qualified human has not reviewed an AI-assisted workflow before producing a client-facing output.

Literacy Program Design: Beyond Awareness

Most corporate AI training programs stop at awareness: employees learn that AI exists, that it carries risks, and that they should consult the policy before using it. Awareness is necessary but insufficient. It does not build the capability to make good decisions in the specific, ambiguous, real-time situations where AI governance actually matters. Literacy programs that go beyond awareness are designed to develop skills and judgment, using instructional methods that require employees to practice the capabilities they are being asked to build.

An effective AI literacy program for enterprise managers covers five content areas. The first is AI capability mapping: what current AI tools can and cannot do reliably, with concrete examples relevant to the manager's function. This area should be updated annually as AI capabilities evolve. The second is governance framework orientation: what the organization's AI tool catalog contains, how to request evaluation of a new tool, and how to report a discovered shadow AI tool. The third is prompt engineering for business use: practical guidance on constructing prompts that produce useful outputs for the specific use cases the manager's team is most likely to encounter. The fourth is output quality assessment: how to evaluate AI-generated content before relying on it, including common failure patterns such as hallucination, overconfidence, and context loss. The fifth is escalation judgment: how to recognize a situation that requires consultation with IT, legal, compliance, or HR before proceeding.

Delivery methods should combine structured learning with immediate application. A ninety-minute instructor-led session covering all five areas, followed by a thirty-day practice period during which managers apply what they learned in their actual work and can access a governance resource for questions, produces better behavioral change than a self-paced module with no follow-through. The practice period should be followed by a brief structured debrief in which managers share one example of AI judgment they exercised during the period. This creates peer accountability, surfaces useful case study material, and reinforces the expectation that AI literacy is a professional skill, not a one-time training event.

Role-Differentiated Curricula

The AI literacy needs of a finance analyst, a software engineer, a clinical coordinator, and a human resources business partner differ, and a single organization-wide AI training curriculum will serve none of them well. Role-differentiated curricula recognize that AI governance knowledge is most effective when it is connected to the specific workflows, risk categories, and decision contexts that are relevant to the employee's function.

Building role-differentiated curricula requires governance teams to partner with functional leaders to map the AI use cases most relevant to each role, identify the governance requirements that apply specifically to those use cases, and design learning content around the scenarios those employees actually encounter. A curriculum for finance team members might focus heavily on AI use in financial modeling, vendor contract review, and regulatory reporting, with specific guidance on which data types are authorized and which review requirements apply to AI-assisted outputs before they are entered into formal records. A curriculum for client-facing professionals might focus on AI in client communication, proposal development, and competitive analysis, with specific guidance on attribution requirements, confidentiality obligations, and quality review standards.

Diagram 8.3: Role-Differentiated AI Literacy Curriculum Map

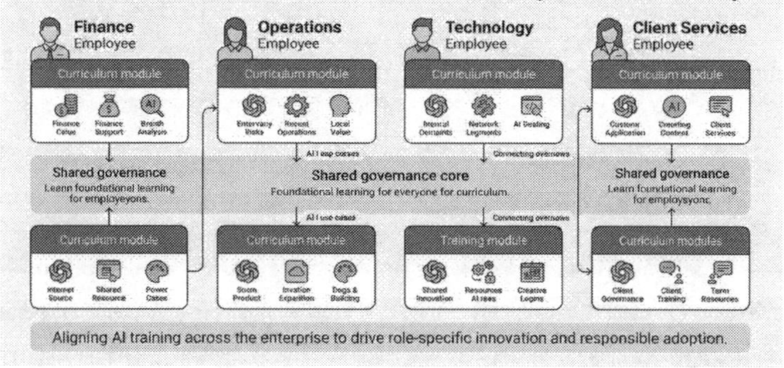

8.4 Building Distributed AI Judgment

The AI Champion Network

No governance team has the capacity to be the primary source of AI guidance for every employee in the organization. Distributed AI judgment requires a network of trained advocates, AI Champions embedded in business units across the enterprise, who can answer practical questions, surface emerging AI use cases, provide peer-to-peer encouragement of compliant use, and serve as early warning channels for governance challenges that are developing at the team level before they become formal incidents.

AI Champions are typically senior individual contributors or early-career managers with demonstrated interest in AI and credibility with their peers. They are not governance enforcers; they are knowledgeable colleagues who have volunteered to deepen their AI expertise and share it with their teams. Their role is explicitly not to make governance decisions. When a question exceeds their knowledge, they escalate to the governance team rather than providing an informal ruling that may be wrong. The value of the Champion network is not decision-making capacity but a significant expansion of the organization's ability to answer everyday AI questions quickly and from a trusted source.

Recruiting AI Champions requires two things: a clear description of what the role involves and what support Champions receive, and active outreach to the managers and senior employees who are most likely already using AI effectively. The governance team should approach this recruitment as they would when identifying internal beta testers for a new technology: they are looking for people who are already engaged, credible, and motivated, not for volunteers who see the Champion designation as a career credential without a genuine interest in the work.

Supporting AI Champions requires ongoing investment from the governance team. Monthly briefings covering policy updates, newly sanctioned tools, recently denied tools, and the rationale, and emerging shadow AI patterns give Champions the current knowledge they need to answer questions accurately. A private communication channel, a shared workspace, a regular call, or a curated newsletter creates a community of practice that allows Champions to share what they are seeing in their business units and learn from each other's experiences. Annual or semi-annual in-person convenings give Champions the relationship foundation that makes peer-to-peer AI guidance feel like a professional community rather than delegated compliance work.

Diagram 8.4: AI Champion Network Architecture

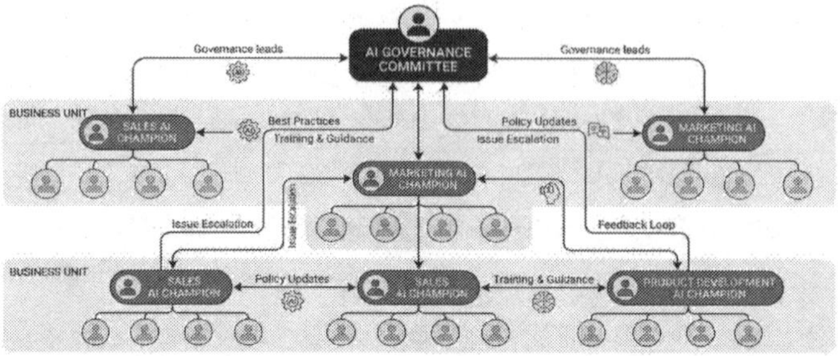

Decentralized AI Champion Network structure facilitates widespread implementation with centralized oversight.

Peer-to-Peer AI Education

Formal training programs reach employees during scheduled time, in structured settings, with standardized content. The AI questions that actually shape behavior arise in the middle of work, when an employee is deciding whether to paste a client document into an AI tool, when a team lead is wondering whether an AI-assisted analysis needs human review before it goes to the board, or when a manager is uncertain whether a new browser extension installed by a team member requires governance notification. Peer-to-peer education addresses these real-time questions in ways formal training cannot.

Peer-to-peer AI education in a governed organization takes several forms. Team-level learning sessions, facilitated by the AI Champion or a manager who has completed advanced AI literacy training, allow small groups to work through realistic use-case scenarios together, practice prompt construction for their specific workflows, and discuss governance questions in the context of their actual work. Show-and-tell formats brief presentations by individual team members on how they used an AI tool, what they learned about its capabilities and limitations, and how they applied the governance framework to normalize AI use as a professional skill and build collective understanding of what responsible use looks like in practice.

The governance team's role in peer-to-peer education is to provide the materials, scenarios, and content that Champions and team leaders use in these sessions, and to review and approve any AI use case examples before they are shared widely. This review serves two purposes: it ensures that the examples describe compliant behavior the governance team is comfortable endorsing, and it identifies case study material to add to the broader internal case study library described earlier in this chapter.

Escalation Confidence

One of the most underappreciated dimensions of distributed AI judgment is what governance researchers call escalation confidence: the employee's belief that bringing a governance question forward will produce a useful outcome rather than a punitive one. Employees who lack confidence in escalation do not ask questions; they make their own determinations, often without the context or expertise to make them well-informed. The result is exactly the pattern that governance programs are trying to prevent:

consequential AI decisions made without appropriate oversight because the people making them were not confident that seeking guidance was safe.

Building escalation confidence requires attention to how governance teams respond to questions. Response time matters: a question that goes unanswered for three weeks trains the employee to stop asking and start deciding unilaterally. Tone matters: responses that lecture, warn, or reference policy consequences before addressing the actual question teach employees that escalation is risky. Resolution quality matters: responses that provide a clear, actionable answer to the specific question asked build the belief that governance teams are useful resources. Governance teams should track question volume as a health metric for escalation confidence. A sudden drop in questions is often a signal that employees have stopped asking, not that they have stopped encountering governance-relevant situations.

8.5 Communication Cadence and Ongoing AI Conversation

Regular Updates People Actually Read

The internal AI governance communication landscape is crowded. Policy updates, training reminders, security alerts, and vendor change notices compete for employee attention alongside other organizational communications. Governance communications that want to be read and, more importantly, to be acted on must be designed with the same intentionality that the best internal communications teams apply to any audience-sensitive message.

The format and frequency of AI governance communications should be driven by the audience's information needs, not by the governance team's reporting preferences. A monthly one-page AI governance digest covering tool catalog changes, recent use case examples, and one governance question answered in plain language is more likely to be read and remembered than a comprehensive quarterly report that requires twenty minutes to review. The digest should have a consistent format so that regular readers know where to find each type of information, and it should lead with something immediately useful to the reader rather than with a governance housekeeping update.

Communications that announce changes to the AI tool catalog deserve particular attention. When a tool is newly sanctioned, the announcement should explain what it does, who can access it, what data types it is authorized to process, and how to get started. When a tool is newly prohibited, the announcement should explain why and what the approved alternative is. When the conditions of an existing tool change, the announcement should explain what changed and what employees should do differently. These communications should be brief, specific, and written in language that a non-technical employee can act on without additional research.

Feedback Mechanisms for Governance Design

The most common failure mode in AI governance communication is treating the employee audience as a recipient rather than a participant. Employees have direct experience of how governance policies interact with their work: they know which approved tools fall short of their needs, which evaluation processes take too long to provide useful answers, and which policy language is ambiguous enough to produce different interpretations across the team. That experience is governance design intelligence, and organizations that do not collect it systematically are designing governance in the dark.

Feedback mechanisms for governance design should be lightweight enough to use regularly and specific enough to produce actionable input. A short quarterly survey, five to seven questions covering tool satisfaction, policy clarity, process friction, and shadow AI awareness, provides a baseline that governance teams can track over time. An always-open suggestion channel, a simple form, or a dedicated email address captures in-the-moment feedback when it is most specific and most actionable. Annual focus groups with representatives from each business unit give the governance team qualitative depth that survey data cannot provide: the reasons behind the ratings, the specific workflow moments where governance either helps or hinders, and the suggestions for improvement that employees are unlikely to put in writing on a form.

Feedback collected through these mechanisms should be reviewed at regular governance team meetings, with specific agenda time dedicated to evaluating whether input warrants a policy change, a tool evaluation priority shift, a process improvement, or a communication clarification. When employee feedback leads to a visible governance change, the change should be announced in a way that acknowledges the source: “Based on feedback from teams across the organization, we have streamlined the tool request process to provide a response within five business days.” That acknowledgment closes the feedback loop and reinforces that participation in governance design produces results.

Diagram 8.5: AI Governance Feedback Loop

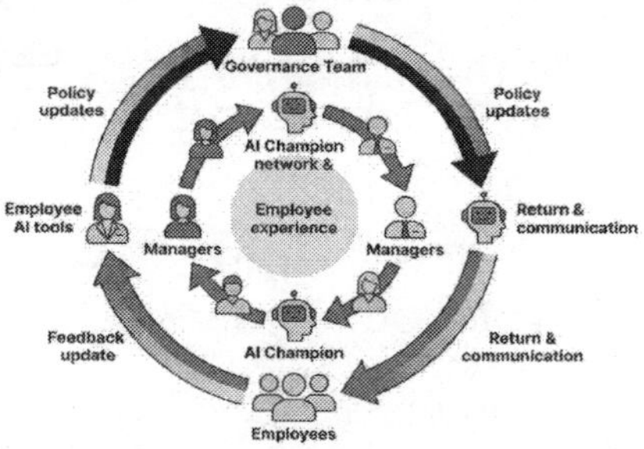

This infographic outlines the feedback loop between governance, managers, and employees regarding the AI experience.

Celebrating Compliant Innovation

Governance programs that only communicate about risks, violations, and requirements create a mental model in which AI governance is synonymous with constraint. Programs that also communicate about compliant AI innovation, how employees use sanctioned tools to do what the organization values, create a mental model in which governance is the infrastructure that enables that innovation. The distinction matters enormously for the program's long-term credibility and adoption.

Celebrating compliant innovation does not require elaborate recognition programs. It requires consistent, specific, public acknowledgment when an employee or team uses AI in a way that is both compliant and impressive. A brief mention in an all-hands meeting, a spotlight in the internal newsletter, a note from the CAO or a senior business leader, or a presentation opportunity at a team or divisional meeting, these are low-cost signals that compliant AI use is professionally valued in the organization. When employees see that responsible AI use is recognized, they are more likely to seek organizational AI literacy resources that help them use AI responsibly, to bring shadow AI tools forward through governance channels, and to view governance as something worth engaging with rather than something to route around.

The governance team should maintain a recognition pipeline: a running list of compliant AI use cases identified through the AI Champion network, the case study collection process, or direct manager nominations. From that pipeline, three to five cases per quarter should be selected for recognition, based on their potential to inspire peer adoption of similar behaviors. The cases should be diverse across function, role level, and type of AI use, showing that compliant AI innovation is possible across the organization, not just in the technically sophisticated corners of it.

Diagram 8.6: Compliant Innovation Recognition Program

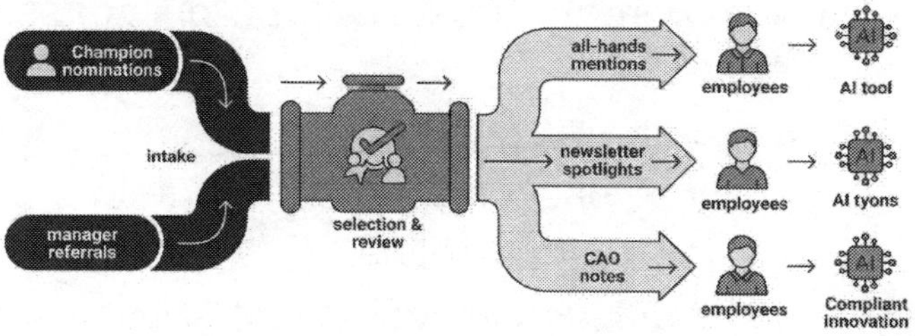

Caption: One compliant innovation and a sen type of compliant innovation being celebration

8.6 Manager's Checklist

- Audit your team's AI communications: count how many lead with risk warnings versus how many lead with what AI makes possible. Rebalance toward enablement framing.
- Complete the role-differentiated AI literacy curriculum for your function before asking your team to complete theirs. Your credibility as a governance model requires demonstrated fluency.
- Nominate at least one member of your team as an AI Champion candidate if your business unit does not already have one.

- Run a team-level AI show-and-tell within the next quarter: ask each team member to share one AI use case from the past month, whether compliant or not, and use the discussion to reinforce governance expectations in context.
- When a team member brings you a governance question, respond within two business days with a clear answer or a named escalation path. Unanswered questions build concealment habits.
- Review the AI governance monthly digest and identify at least one item per issue to discuss in your next team meeting.
- Submit at least one compliant AI use case from your team to the case study collection process each quarter.
- Participate in the governance feedback survey and encourage your team to complete it. Their input shapes the tools and policies you work with.
- When a team member uses AI in a way that is both compliant and impressive, acknowledge it specifically and publicly within the team.
- When you communicate about AI governance to your team, lead with what AI enables before addressing what governance requires.

8.7 What This Chapter Established

The enablement pivot is not a communications strategy layered on top of an unchanged governance model. It is a fundamental redesign of how the organization positions AI governance in the experience of the people subject to it. That redesign touches leadership messaging, training design, peer education infrastructure, communication formats, and recognition practices, all of the organizational mechanisms through which culture is shaped rather than declared.

Organizations that make the enablement pivot successfully develop a self-reinforcing dynamic: compliant AI use is recognized and shared, which motivates more compliant AI use, which produces more case study material for communication, which reaches more employees through more credible channels, which builds more AI literacy, which produces better AI judgment, which reduces shadow AI risk. That dynamic is the opposite of the restriction trap, in which governance communications generate anxiety rather than capability, and employees route around governance rather than through it.

The practical tools for making this pivot, leadership messaging frameworks, role-differentiated curricula, AI Champion networks, feedback mechanisms, and recognition programs, are not theoretically complex. They require sustained organizational will and genuine investment in the people they serve. The final chapter of this book addresses the structural and operational requirements for sustaining that investment over time, as AI capabilities, organizational pressures, and governance challenges continue to evolve.

9 Sustained Visibility

The governance program at a state government agency had taken two years to build. The AI acceptable use policy had been drafted with input from IT, legal, HR, and the agency's privacy office. The approved tool catalog included 12 tools across four use-case categories. The AI Champion network had fourteen members embedded across the agency's eight divisions. Quarterly audits had been conducted for six consecutive quarters, and the discovery rate for unsanctioned tools had declined from seventeen incidents in the first quarter to four in the most recent one. By any reasonable measure, the program was working.

Then the agency's primary productivity suite vendor released a major update. The update bundled an AI assistant directly into the email and document applications that every agency employee used daily. The AI assistant was enabled by default. It had access to the employee's email inbox, calendar, and draft documents. It could generate replies, summarize email threads, and suggest edits to in-progress documents. Within seventy-two hours of the update, the AI assistant was actively in use by several hundred employees who had not made any deliberate choice to adopt it, who had not received any training on its data handling implications, and whose use of it fell entirely outside the governance framework the agency had spent two years constructing.

The governance program had been built to manage deliberate shadow AI adoption. The vendor update had created instant, ambient, invisible AI adoption at scale — a qualitatively different challenge that the existing framework was not designed to address. The agency's governance team had to decide, under pressure and without the benefit of an established process, how to respond to an AI deployment they had not anticipated, in a tool they had fully sanctioned, with a capability they had never evaluated.

That scenario — the unexpected emergence of AI in an already-approved context — is one of the defining challenges of sustained AI governance. This final chapter addresses the operational requirements of keeping a governance program functional, current, and mission-aligned as the AI landscape continues to evolve. The work described here is not a phase of the governance lifecycle that comes after the hard work is done. It is the hard work, repeated continuously as the environment changes.

9.1 Why Sustained Visibility Is the Hardest Part

Launching an AI governance program is a bounded project with a visible endpoint. Writing the policy, building the registry, standing up the Champion network, completing the first round of tool evaluations — these are achievable milestones that produce concrete artifacts and visible organizational progress.

Sustaining the program is a different kind of work: it has no endpoint, its achievements are largely preventive and therefore invisible, and its primary adversary is not employee resistance or vendor non-compliance but organizational inertia. Governance programs that are not actively maintained deteriorate. Policies become dated without a revision process. Registries accumulate stale entries as tools change and teams evolve. Champion networks lose their effectiveness as members change roles and the briefing cadence slips. Evaluation queues grow longer as new AI capabilities emerge faster than the governance process can assess them.

The pace of AI development poses a specific sustainability challenge with no parallel in previous governance domains. The AI tool landscape in 2025 looks categorically different from that in 2023 — and the landscape in 2027 will look different again. Each wave of new capability brings new shadow AI adoption vectors, new vendor terms that require re-evaluation, and new use cases that the existing policy may not have anticipated. A governance program built around the AI tools of its launch year will be meaningfully out of date within 18 to 24 months, not because the governance team failed, but because the environment changed faster than annual policy cycles can track.

Sustained visibility is the organizational commitment to staying current: maintaining detection capabilities that keep pace with adoption patterns, running governance review cycles that respond to environmental change rather than calendar-driven cycles, and building the organizational competencies that enable the governance team to understand and respond to new AI capabilities before they become governance crises. This chapter provides the operational architecture for that commitment.

DIAGRAM 9.1: GOVERNANCE DECAY WITHOUT SUSTAINED MAINTENANCE

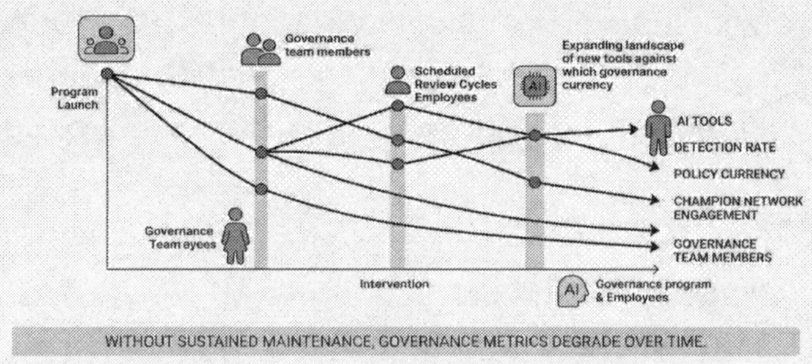

9.2 Continuous Discovery in a Persistent Landscape

Automated Detection Mechanisms

The detection architecture established in Chapter 4 — DNS log analysis, web proxy monitoring, expense report mining, and employee disclosure programs — provides the foundation for ongoing discovery. Sustaining that architecture requires two forms of maintenance: technical upkeep to keep detection tools current with the evolving landscape of AI services, and process upkeep to ensure detection outputs are reviewed and acted on within defined timeframes.

Technical maintenance of automated detection mechanisms focuses primarily on keeping watchlists up to date. DNS filtering and web proxy monitoring operate against lists of domains associated with AI services. Those lists require regular expansion as new AI products emerge. The governance team should subscribe to AI industry tracking resources — analyst reports, product release newsletters, and technology press — to stay up to date on new consumer and enterprise AI tools entering the market. It should add their associated domains to the monitoring watchlist within thirty days of public availability. This requires a designated team member whose responsibilities include tracking the AI landscape, not merely governance administration.

Process maintenance of automated detection mechanisms requires defining and enforcing response timelines for detection outputs. A DNS alert that identifies an employee accessing a new AI service should trigger a defined workflow: a notification to the manager, a brief structured conversation following the disclosure framework in Chapter 4, an entry in the AI tool registry as a discovered unsanctioned tool, and initiation of the evaluation process in Chapter 7. Without defined response timelines and accountability, detection outputs accumulate as unactioned alerts, and the organization has the appearance of a detection capability without the governance substance.

Quarterly AI Tool Audits

In addition to continuous detection mechanisms, sustained governance requires periodic structured audits of the AI tool environment. A quarterly audit serves three purposes that continuous detection cannot: it provides a systematic review of the full tool registry to identify entries that require update, re-evaluation, or retirement; it produces a point-in-time picture of the organization's AI tool footprint that can be compared across quarters to identify trends; and it creates a regular organizational touchpoint for AI governance that reinforces the program's ongoing relevance.

A quarterly audit covers five areas. Registry currency review examines each tool in the approved catalog to confirm that it is still in active use, that the vendor terms and security posture have not materially changed, and that the authorized use cases remain current with how the tool is actually used. Conditional approval review identifies all tools with time-boxed authorizations and confirms that review dates are being met or, if they have passed, that a disposition has been recorded. Vendor change monitoring reviews changes in vendor terms, ownership, data handling policies, or security certifications for all approved tools since the previous audit. Shadow AI intake reviews the queue of discovered unsanctioned tools to ensure that all are either under active evaluation or have a recorded disposition. Emerging tool review identifies AI tools that the organization has not yet evaluated but are gaining adoption among employees or peer organizations in the same industry.

The quarterly audit output should be a brief, structured report that the AI steering committee reviews at its next scheduled meeting. The report should not be exhaustive — it should highlight material changes, unresolved evaluation items, and trends that require governance team attention. Audit reports that are comprehensive but unused are governance overhead without governance value.

Diagram 9.2: Quarterly Audit Cycle and Outputs

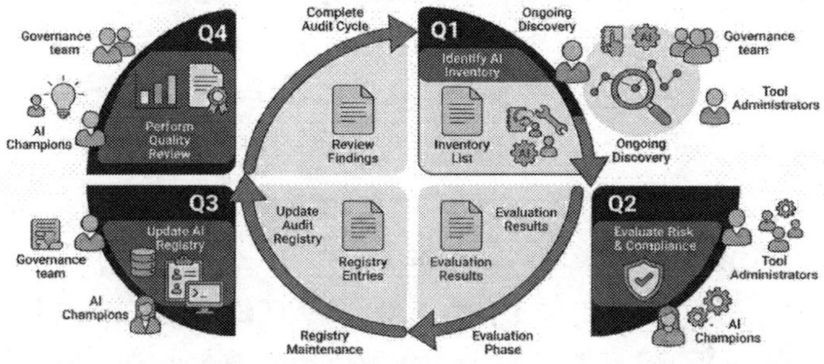

This recurring quarterly process ensures continuous oversight, compliance monitoring, and accurate documentation for all organizational AI systems.

Signal Monitoring: Tracking Adoption Patterns

Detection mechanisms identify specific tool use. Signal monitoring identifies adoption patterns and the organizational dynamics that predict where shadow AI is likely to emerge before specific instances are detected. Signal monitoring is a forward-looking complement to the backward-looking function of detection: it helps governance teams anticipate governance challenges rather than merely respond to them.

Adoption pattern signals include several categories. Productivity pressure indicators such as headcount reductions, increased output targets, or tightened delivery timelines in specific business units predict increased pressure toward AI workarounds. They should trigger proactive outreach from the governance team to affected managers. Vendor market signals new AI tool releases, significant product updates to existing tools, or major media coverage of a new AI capability, predict increased employee awareness and trial behavior, and should trigger prioritization of the evaluation queue. Internal request patterns cluster similar tool evaluation requests within the same business unit, signaling either a team-level adoption pattern or a coordinated effort to get a specific tool approved, both of which warrant direct engagement with the relevant manager. Network traffic anomalies, sudden spikes in traffic to AI-associated domains from specific network segments, may indicate coordinated team-level adoption of a new tool that has not yet surfaced through other detection channels.

Signal monitoring does not require sophisticated technology. It requires a governance team member with designated responsibility for environmental scanning, a defined set of signal categories to monitor, a process for translating signal observations into governance actions, and a communication channel between the signal monitor and the team member responsible for the evaluation queue. The technology supporting signal monitoring is largely the same technology already in use for detection: web proxy logs, DNS monitoring, expense report review, and direct input from the AI Champion network.

9.3 Governance Model Evolution

The Governance Review Calendar

AI governance policies and frameworks should not be static documents. They should be living instruments that are reviewed and revised on a defined schedule, with trigger-based provisions that allow out-of-cycle revision when the environment demands it. Establishing a governance review calendar, a documented schedule of when each governance component is reviewed and by whom, is the mechanism that turns the intent to stay current into an organizational commitment with accountability.

A practical governance review calendar covers four timeframes. Annual reviews address the full acceptable use policy, the AI tool catalog's classification criteria, the risk scoring framework used in tool evaluation, and the AI Champion program's design and membership. These are the foundational governance documents that require a comprehensive review but do not change with every development in the AI landscape. Semi-annual reviews address the AI Champion training curriculum, the role-differentiated literacy program content, and the onboarding checklist for newly sanctioned tools. These components require more frequent revision because the AI capabilities and tools they reference change faster than annual cycles can track. Quarterly reviews address the tool registry, the evaluation queue status, the list of domains included in detection monitoring, and the audit outputs described in the previous section. Monthly reviews address the governance communications calendar, the case study pipeline, and the recognition program output—the operational components that require frequent adjustments to remain effective.

Trigger-based revision provisions allow the governance team to initiate out-of-cycle review when a specific event warrants it: a major regulatory development affecting AI use in the organization's industry, a significant security incident involving an AI tool at a peer organization, a material change in the terms of a broadly used approved tool, or the emergence of a new AI capability category that the existing policy did not anticipate. Trigger-based revision is not an invitation to perpetual policy instability; it is a structured mechanism for responding to genuine material changes without waiting for the next scheduled review cycle.

Technology Horizon Scanning

AI governance programs that are built around current capabilities will always be behind. The governance team's ability to stay relevant depends on its capacity to understand emerging AI capabilities before they appear in the organization's shadow AI detection logs. Technology horizon scanning is the systematic practice of monitoring the AI development landscape to identify capabilities that will be relevant to the organization's governance framework over the next 12 to 24 months.

Horizon scanning for AI governance focuses on three categories of developments. New tool categories are emerging AI product types that do not fit within the use-case classifications in the current policy framework. Examples from recent years include multimodal AI tools that process images and documents together, AI-powered browser agents that take actions on behalf of users across multiple web services, and AI coding assistants that generate and execute code with minimal human supervision. Each of these categories raised governance questions that existing policies had not addressed, and organizations that first encountered them through shadow AI discovery were at a disadvantage relative to those that had anticipated them. Second-order capability developments are features added to existing approved tools that change their risk profile without changing the tool's primary classification. An AI assistant added to an approved document management system, a summarization feature added to an approved meeting platform, or a predictive text capability added to an approved email client each represents a governance challenge embedded in an already-approved tool. Regulatory technology signals are AI-specific regulatory developments in the organization's jurisdiction or sector that indicate where compliance obligations related to AI use are heading, allowing the governance team to build readiness before requirements become enforceable.

Horizon scanning does not require the governance team to become technology analysts. It requires designated attention time allocated to reading the AI technology and policy landscape and a structured process for converting observations into governance planning inputs. A brief quarterly technology landscape briefing, prepared by the team member responsible for horizon scanning and reviewed at the AI steering committee meeting, keeps the full governance committee informed without requiring each member to conduct their own monitoring.

Diagram 9.3: Technology Horizon Scanning Framework

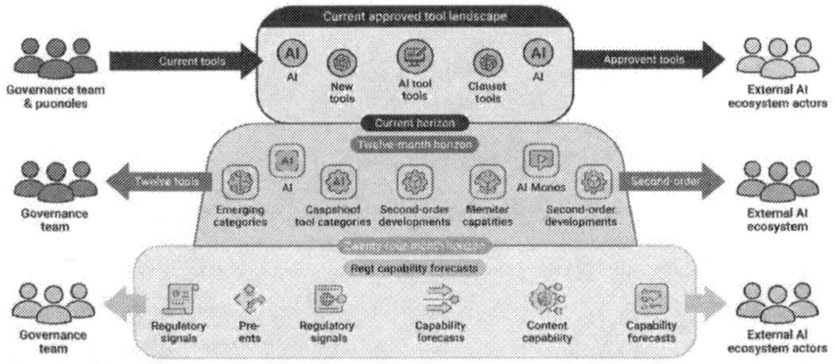

Technology horizon scanning: exampls to illustratet the expanding surface of **governance-relevant** developments.

Stress Testing Against Emerging AI

Horizon scanning identifies what is coming. Stress testing determines whether the governance framework can handle it. Stress testing is the practice of evaluating existing governance policies and processes against emerging AI capabilities to identify gaps before those capabilities arrive in the organization's environment. It is a proactive governance investment that converts horizon scanning intelligence into policy readiness.

A governance stress test for an emerging AI capability follows a structured sequence. First, describe the capability in operational terms: what does it do, who is likely to want to use it, what data does it process, what outputs does it produce, and what decisions might it influence? Second, evaluate the current policy against that description: does the acceptable use framework cover this capability, or does it fall into a gap? Are the data handling requirements in the current policy sufficient for the risk profile this capability presents? Does the evaluation framework in Chapter 7 provide adequate criteria for assessing a tool with this capability? Third, identify the gaps: where is the current framework insufficient? Fourth, determine remediation: should the gap be addressed through a policy amendment, an addition to the evaluation criteria, a new use case category in the tool catalog, or a new training requirement for the literacy program? Fifth, prioritize: given the estimated timeline for this capability to reach the organization, when must remediation be completed to maintain governance readiness?

Stress testing should be conducted for at least two or three emerging capability categories per year, selected by the governance team based on horizon-scanning observations. The outputs of stress testing identified gaps and remediation priorities that feed directly into the annual policy review cycle, ensuring that the review is driven by actual governance needs rather than by a calendar-driven exercise in superficial revision.

9.4 Preparing for Agentic and Embedded AI

Autonomous AI and New Accountability Questions

The AI governance challenges addressed throughout this book, such as shadow adoption, prompt-level data exposure, output quality assurance, and vendor assessment, are primarily challenges associated with AI tools that assist human workers. The human decides when to use the tool, what to submit to it, and how to use its outputs. The AI is a capability in the human's workflow, not an actor in its own right. Agentic AI fundamentally changes that relationship, and the governance implications are significant enough to warrant dedicated attention even in governance frameworks that are not yet encountering agentic AI in their organizations.

Agentic AI refers to AI systems that take sequences of actions autonomously to accomplish multi-step objectives, without requiring human approval at each step. An agentic AI system might browse the web to gather research, draft a document, send a communication, update a database record, and schedule a follow-up meeting, all in response to a single high-level instruction, and all without human review of each action. The governance questions raised by agentic AI are categorically different from those raised by AI assistants. When an agentic system sends an email, who authorized that communication? When it accesses a vendor's website to gather competitive intelligence, does that constitute a data access that requires disclosure? When an enterprise system updates a record, does that update follow the same audit trail requirements as a human-initiated update?

Governance frameworks that have not addressed agentic AI will face a specific readiness gap when employees begin using agentic AI tools, and employee adoption of consumer-grade agentic AI tools is already underway. The acceptable use framework must address how agentic AI is classified, what authorization is required before an agentic system is permitted to act on behalf of an employee, what human-review checkpoints must be built into agentic workflows, and what audit-trail requirements apply to actions taken by agentic systems in enterprise contexts.

AI in Critical Workflows

As AI capabilities mature, the boundary between AI-assisted workflows and AI-determined outcomes is eroding in consequential ways. Early enterprise AI use concentrated in relatively low-stakes tasks: drafting first versions of routine documents, summarizing long reports, and generating initial code stubs for review by a senior engineer. As organizations develop greater confidence in AI outputs and AI capabilities improve, AI is being used in workflows with higher stakes, such as credit assessment, clinical triage support, hiring recommendations, regulatory compliance determinations, and supply chain decision support.

The governance implications of AI in critical workflows are distinct from the implications of AI in productivity support tasks. In productivity support, a human reviews the AI output and applies judgment before the output has any organizational consequences. In critical workflows, the gap between AI output and organizational consequences is sometimes narrowed or eliminated in fully automated decision pipelines. The governance framework must be able to distinguish between these contexts and apply proportionate controls. Critical workflow AI requires human review, defined escalation thresholds, audit-trail specifications, performance-monitoring obligations, and periodic outcome review to assess whether the AI's contributions produce fair, accurate, and mission-aligned results.

A critical workflow designation process should be part of the tool evaluation framework. When a tool evaluation is initiated, one of the early evaluation steps should be a workflow-criticality assessment: Does the tool's intended use case involve decisions with significant consequences for individuals, regulatory compliance status, financial commitments, or client relationships? Tools that pass the criticality threshold should be subject to an extended evaluation process with additional review requirements, and their sanctioning should specify the human review and audit trail obligations that apply to their use in critical contexts.

Diagram 9.4: Critical Workflow AI Governance Requirements

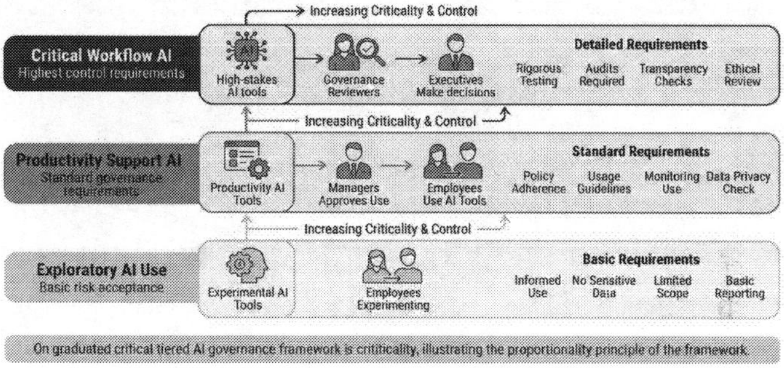

Governance Architecture for Autonomous Systems

When the governance framework must address autonomous AI systems that take actions rather than merely generate content, the core governance design principles remain the same, but their application requires adaptation. Authorization must be prospective: the organization must define in advance what categories of actions an agentic system is permitted to take, rather than evaluating individual outputs after the fact. Monitoring must be action-level, not just output-level: the governance record must capture what actions the system took, not merely what it produced. Human oversight must be structural: the workflow design must include defined checkpoints at which a human reviews the system's actions before those actions cross certain consequence thresholds, rather than relying on ad hoc human judgment about when to intervene.

The governance architecture for autonomous AI systems should address six elements. Authorization scope defines the categories of actions the system is permitted to take and the contexts in which it is authorized to operate. Action logging specifies what must be recorded about each action the system takes, at what level of detail, and for how long. Human review checkpoints identify the consequence thresholds that trigger mandatory human review before an action is completed. Rollback provisions specify whether and how actions taken by the system can be reversed if a review determines they were incorrect or unauthorized. Incident classification defines what constitutes an incident for an autonomous system: not just security failures, but also material errors, unauthorized actions, or outputs that cross defined harm thresholds. Performance monitoring establishes how the system's output quality and decision accuracy are tracked over time, and what remediation or suspension triggers apply.

9.5 The AI Governance Maturity Model

Five Stages From Reactive to Anticipatory

Organizations at different stages of AI governance development have different operational priorities. A governance maturity model provides a structured way to assess an organization's current state and what the next stage of development requires. The five-stage model presented here describes a progression from reactive governance, which responds to shadow AI incidents as they are discovered, to anticipatory governance, which operates ahead of the AI development curve through active horizon scanning, stress testing, and proactive policy development.

Stage one is reactive governance. At this stage, the organization has no formal AI governance framework. Shadow AI is discovered through incidents rather than through proactive detection. Responses are ad hoc and inconsistent, driven by whoever first encounters the issue rather than by a defined process. Policy, if it exists, is a generic acceptable use statement that does not address AI specifically. Most organizations are at this stage when they first recognize that AI governance is a material concern.

Stage two is foundational governance. The organization has a formal AI acceptable use policy, a basic approved tool registry, and defined processes for tool evaluation and employee disclosure. Detection is primarily reactive, driven by incidents and disclosure rather than by automated monitoring, but the framework exists to respond consistently when shadow AI is discovered. Most of the governance infrastructure described in this book's early chapters has been established, but the program remains primarily driven by discovery events rather than by systematic monitoring.

Stage three is operational governance. The organization runs systematic detection, quarterly audits, and a living tool registry. The AI Champion network is active, the literacy program is operational, and the tool evaluation process runs within defined timelines. The governance framework addresses most shadow AI challenges that arise, though it may still be caught off guard by new capability categories. The organization is compliant with its stated governance commitments and can demonstrate that compliance to auditors and regulators.

Stage four is adaptive governance. The organization has added horizon scanning, governance stress testing, and trigger-based policy revision to its operational program. It responds to new AI capability categories before they produce governance incidents rather than after. The governance framework is regularly tested against emerging capabilities and updated proactively. The governance team has sufficient expertise in AI technology to engage meaningfully with the vendor assessments and capability evaluations required by the program.

Stage five is anticipatory governance. The organization operates at the frontier of AI governance practice: contributing to industry governance standards, sharing best practices with peer organizations, and actively shaping the regulatory environment in which enterprise AI governance operates. The governance program is mission-aligned at the strategic level — not merely compliant at the operational level and it has the organizational credibility and leadership support to evolve ahead of the AI development curve rather than behind it.

Diagram 9.5: AI Governance Maturity Model

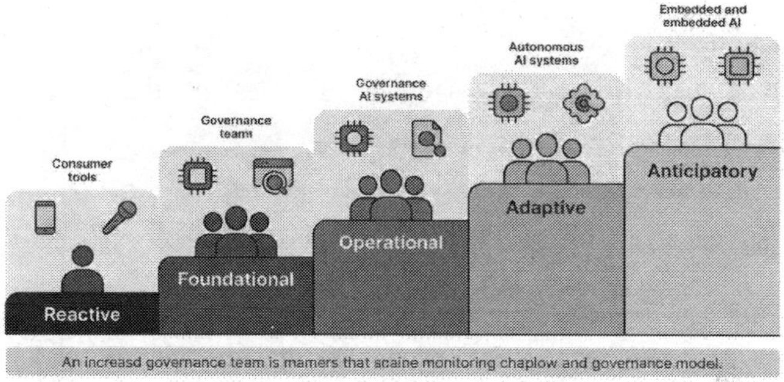

Maturity-Indexed Priorities

Each maturity stage has a distinct set of governance priorities that should drive the program's resource allocation and leadership attention. Understanding which stage the organization occupies is the starting point for sound governance investment decisions, because the returns on governance investment are highest when resources are directed to the stage-appropriate priority rather than to capabilities the organization is not yet ready to use.

For organizations at stage one, the priority is to establish the framework: drafting a basic AI-acceptable use policy, building the initial approved tool registry, and establishing disclosure and evaluation processes that enable the organization to respond consistently to shadow AI discoveries. Leadership investment at this stage goes into governance infrastructure: policy development, registry technology, and training resources. For organizations at stage two, the priority is process operationalization: moving from a governance framework on paper to one consistently applied in practice. This requires investment in the AI Champion network, the literacy program, and the communication cadence that keeps the governance framework visible and relevant. For organizations at stage three, the priority is detection and monitoring: building automated, periodic discovery capabilities that reduce reliance on disclosure-driven governance. This requires investment in technical detection tools, quarterly audit processes, and the analytics capabilities to make sense of detection outputs.

For organizations at stage four, the priority is horizon scanning and stress testing: building the organizational capacity to anticipate governance challenges before they arrive. This requires investment in designated analytical capacity, relationships with external AI governance communities, and a governance committee with the sophistication to engage with emerging questions about AI capabilities. For organizations at stage five, the priority is leadership: contributing the organization's governance knowledge to the broader ecosystem through publications, standards participation, and peer engagement, while maintaining the internal program at the frontier of practice.

Measuring Progress: Governance KPIs

AI governance programs that cannot measure their own progress cannot demonstrate their value or make defensible resource allocation decisions. Governance KPIs provide the metrics that allow the program to track its effectiveness, identify deterioration before it becomes a crisis, and communicate its contribution to organizational leadership in terms that connect to business outcomes.

A core set of governance KPIs covers five dimensions. Detection effectiveness measures the percentage of shadow AI tools discovered through proactive detection mechanisms, such as DNS monitoring, expense report mining, and network anomaly detection, versus reactive disclosure after an incident. As the program matures, this percentage should increase: a high reactive discovery rate indicates that proactive detection is underperforming. Evaluation velocity measures the average time from tool discovery to disposition: provisional approval, conditional approval, full sanctioning, or prohibition. This metric tracks whether the evaluation process is running at the speed the organization needs to compete with unsanctioned adoption. Policy currency measures the percentage of governance policy documents that are within their scheduled review cycle, a simple but important indicator of whether the governance review calendar is being followed. Literacy penetration measures the percentage of employees who have completed the role-differentiated AI literacy program for their function, and the percentage of managers who have completed the advanced literacy curriculum. Champion network health measures the percentage of business units with an active AI Champion, the frequency with which Champions engage with the governance team's monthly briefings, and the volume of escalations routed through the Champion network rather than directly to governance team members.

These KPIs should be reviewed at each quarterly audit and reported to the AI steering committee. Trending data on how each metric has moved over the past four quarters is more informative than point-in-time snapshots, because it shows whether the program is improving, plateauing, or deteriorating. Governance programs that report only on activity, the number of tools evaluated, the number of trainings completed, and the number of policy documents reviewed, without connecting those activities to outcome metrics, fail to demonstrate whether the activity is producing the governance result the organization needs.

Diagram 9.6: Governance KPI Dashboard

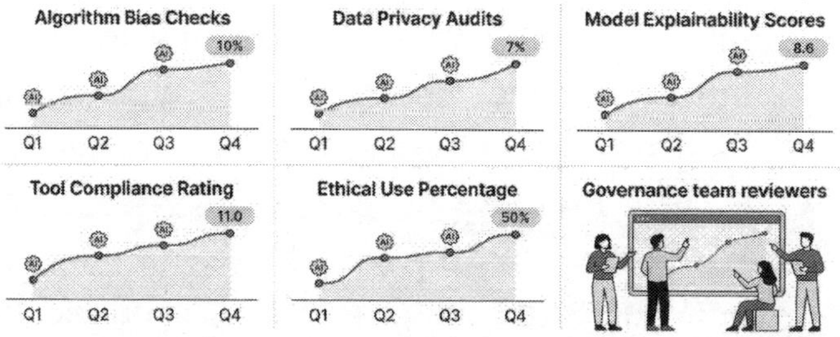

This dashboard tracks key performance indicators for managing AI and algorithmic governance.

9.6 Manager's Checklist

- Subscribe your team to the quarterly AI governance digest and include one governance item from each issue in your next team meeting. Consistency builds the habit of governance awareness.
- Know which AI tools your team is using and whether each is in the approved catalog, under evaluation, or outside the governance framework. This is a foundational management responsibility, not a delegation to IT.
- When your team's productivity environment changes, a software update, a new vendor feature, or a platform migration, check whether the change introduces new AI capabilities that require governance review.
- Participate in the quarterly audit process by providing accurate information about your team's AI tool usage and flagging any tools your team has been using that are not in the registry.

- When you encounter an AI capability that your governance framework has not addressed, raise it through the AI Champion or directly to the governance team. These observations are the raw material of horizon scanning.
- Ensure that any AI use in critical workflows on your team has the human review checkpoints and audit trail documentation required by your governance framework.
- Review your team's AI literacy completion status quarterly and address gaps in literacy, as incomplete literacy creates uneven governance judgment in the team, which creates exposure.
- Know your organization's governance maturity stage and understand what the next stage requires your team's cooperation for the program to advance.
- When the governance team requests your input for the stress testing process, engaging your knowledge of your team's workflows substantively is essential for identifying where emerging AI capabilities create governance gaps.
- Treat AI governance as a recurring management discipline, not a one-time compliance task. Schedule quarterly AI governance reviews on your calendar alongside your other recurring management responsibilities.

9.7 What This Chapter Established

Sustained visibility is not a phase of AI governance; it is the condition under which AI governance actually works. The infrastructure built in the program's early stages, the policy, the registry, the Champion network, and the evaluation process depreciate without maintenance. New AI capabilities create governance gaps. Vendor updates change the risk profile of approved tools. Organizational pressures produce new patterns of shadow AI adoption. The governance team that is not actively maintaining its detection capabilities, reviewing its policy currency, and scanning the technology horizon is governing yesterday's AI environment. At the same time, the organization operates in today's world.

The governance maturity model provides a framework for understanding where the organization stands and what the next stage of development requires. Maturity-indexed priorities ensure that governance investment is directed at the capabilities the organization is ready to build, rather than at sophisticated governance instruments that require a more advanced foundation. Governance KPIs make the program's progress visible and its value demonstrable — both to the governance team itself and to the leadership that funds and supports it.

The arc of this book, from the invisible deployment of Chapter 1 to the anticipatory governance described here, describes a journey that no organization completes in a single cycle. Shadow AI governance is iterative, responsive, and never finished. The organizations that manage it best are those that have accepted that reality and built the operational discipline to stay current despite it. They have governance teams with designated environmental scanning responsibilities, review calendars with real accountability, evaluation processes that move at the speed of adoption, and communication programs that make governance feel like an enabler of the organization's AI ambitions rather than a brake on them. That combination of operational rigor, technical currency, and genuine enablement intent is what responsible AI governance by design looks like in practice.

10 The Organization That Sees What It Is Actually Using

You started this book with a recognition that most senior leaders arrive at reluctantly: your organization adopted artificial intelligence without a procurement decision, a risk assessment, or a governance framework. Employees across every function found tools that made them faster and used them. The nine chapters you have just completed provide a systematic path from that ungoverned starting condition to an operating model in which the enterprise can see what is being used, manage the risks it creates, and sustain that visibility as the AI landscape continues to evolve.

The journey moved from understanding why shadow AI adoption was not a failure of compliance but a predictable response to accessible technology, through the risk landscape that unsanctioned tools create, into the operational challenge of discovering what the organization does not know it is using. You then built the governance architecture policy that people will follow, accountability shared across IT, legal, compliance, and human resources, and the transition pathways that bring rogue tools under recognized management without destroying the productivity they enabled. The final chapters reframed the organizational posture from prohibition to enablement, which is the only stance that prevents the next wave of unsanctioned adoption from replicating the current one.

The call to action is not to ban AI tools or to write a more restrictive acceptable use policy. Both of those responses have already failed in organizations that tried them. The call to action is to build the governance infrastructure that makes unsanctioned intelligence unnecessary by providing sanctioned alternatives that are fast enough, capable enough, and accessible enough that employees stop reaching outside the enterprise to get their work done. That infrastructure requires investment, cross-functional coordination, and sustained executive attention.

Launch the discovery process this quarter. Identify the five tools with the widest unsanctioned adoption. Convene the cross-functional governance team with representation from every function that owns a piece of the risk. Publish the first version of an AI acceptable use policy that your workforce helped design rather than one that was imposed without their input. Designate the first set of sanctioned alternatives and make them available without friction. These are concrete, sequenced actions that move the organization from awareness to governance.

Your workforce has already decided that AI is useful. They are not going to stop using it because a policy says so. The question is whether your organization builds the structure that channels that adoption into something it can see, manage, and defend, or whether it continues to operate with an AI footprint it cannot account for. This book gave you the blueprint. What you build with it starts now.

References

Made in the USA
Monee, IL
10 November 2026

edac9684-1936-48ab-9b79-59a92b3bd5b7R01